# Eugenics and Other Evils

An Argument Against the Scientifically Organized State

by

## G. K. Chesterton

With Additional Articles by his
Eugenic and Birth Control Opponents
Including
Francis Galton, C. W. Saleeby and Marie Stopes
As Well as Articles from
*Eugenics Review* and *Birth Control News*

Edited by

## Michael W. Perry

Inkling Books        Seattle        2000

**Library Cataloging Data**

Chesterton, G. K. [Gilbert Keith] (1874–1936)

*Eugenics and Other Evils: An Argument Against the Scientifically Organized State*

Editor: Perry, Michael W. [Wiley] (1948– )

Other Authors: Francis Galton, C. W. Saleeby, Marie Stopes, *Eugenics Review, Birth Control News*

179 p. 23 cm.

Includes: index, 1 illustration and 9 appendices

ISBN 1-58742-002-3 (pbk.: alk paper)

LC 00-105122

Keywords: Eugenics, Birth control, Sterilization, Science, Bureaucracy, Socialism, Capitalism

HQ755 .C48

575.1 C42

Inkling Books, Seattle, WA Internet: http://www.InklingBooks.com/

Published in the United States of America on acid-free paper

First Edition, Second Printing: March 2001

# Contents

# Foreword

If you haven't already, you'll soon discover that Chesterton was a marvelous writer and that this book ranks among his masterpieces. When it first came out, his foes were forced to concede, through clenched teeth, that in him they faced an opponent who knew how to use pen and humor with great skill. Even today we laugh when we read his blast at foolish laws that treat an ordinary Englishman "as harshly as a thief, and almost as harshly as an honest journalist." Chesterton is that most valuable of all literary friends, a truly honest journalist.

Why have I chosen to devote a quarter of this book to the writings of his opponents? Because I want Chesterton to get proper credit for his great achievement. He wrote in the heat of battle, when the debate over eugenics was at its fiercest. And it is easy to suspect that he wrote much of this book in drafty rail stations rather in the warmth of a well-equipped study. Yet no scholar alive has done as much to expose that strange blend of silliness, scientific bigotry and politically correct arrogance which once went by the proud name of eugenics. It will also help you understand why, when Chesterton suggests to eugenists that they study mixtures of races, he is taunting their bigotry and pretensions of racial superiority.

Even more important, as a writer he stood virtually alone against a juggernaut that threatened to sweep all before it. In 1924, a eugenist could speak proudly of a scholarly bibliography listing thousands of articles on eugenic-related topics (see Appendix D). Today, eugenics has few open friends (though many secret admirers). But when Chesterton wrote its ranks were a virtually 'who's who' of the respectable and powerful. Apart from him, almost no one of importance spoke out against it. If you doubt that, visit any university library and look for other book-length criticisms of eugenics from his era.

When two friends of yours marry or when they become proud parents, you should thank Chesterton that neither of those wonderful events required the approval of experts. Think I am exaggerating? Then read this book's nine appendices. And if you feel that nothing they were advocating would have become law, look at the United States during the 1920s. In 1924, our immigration laws were revised to exclude from our shores the very groups that birth controllers and eugenists were blasting as biologically inferior—Eastern and Southern Europeans (particularly Jews and Italian Catholics). And in 1927, to deal with an alleged threat from those already here, the U.S. Supreme Court, by a vote of eight to one, ruled that forced sterilization was constitutional. Much like Chesterton, the Court's lone dissenter, Pierce Butler, had a well-established reputation as a "reactionary."

Who supported eugenics? Chesterton got it right. They were those whose "creed is the great but disputed system of thought which began with Evolution and has ended in Eugenics." Eugenics was confined almost exclusively to those who took evolution seriously and drew conclusions from it about how society should be structured. They had no religion of their own and made a substitute out of science. Two close relatives of Charles Darwin founded and led the eugenic movement and Darwin himself supported it in private.

The link Chesterton sees between capitalism's 'malefactors of great wealth' and eugenics is also undeniable. In the U.S. the Harriman (rail), Carnegie (steel) and Rockefeller (oil) fortunes generously funded eugenic research and regarded it as one of their pet causes. As Chesterton shrewdly points out, theirs is a dismal world where people are little more than tools, to be manufactured or discarded as the need arises. Eugenics slides easily into that sort of mind.

In keeping with the "other evils" of the title, Chesterton also delivered telling blows against socialism and mentions some in its ranks (George Bernard Shaw and H.G. Wells) as well-known eugenists. But in comparison to capitalists, he doesn't delve deeply into the rationale for their support. The reason is political. The roots of eugenics lie in both Darwinian evolution and in earlier theories about population by Thomas Malthus. Charles Darwin caused no controversy in leftist circles. Karl Marx even wanted to dedicate *Das Kapital* to him. (Darwin refused.) But Malthus was a red flag to the rank-and-file left. His theory began as a direct attack on all schemes that claim to create an ideal world. Go ahead, Malthus said, reorganize society so all are well-fed. In the long run you do no good. Deprived of a high death rate, the population will grow to a point where the land will no longer provide enough food, and you'll have more people starving than before. This made the average socialist suspicious of anyone claiming that the answer to his problems lay in fewer children. That, however, wasn't the attitude of the elite left.

To understand the left's attitude toward eugenics, you must grasp the distinction between the inner and outer rings of an organization. C. S. Lewis described it well in his 1944 University of London speech on "The Inner Ring," and illustrated it in his novel, *That Hideous Strength.* The inner ring of an organization (such as the Fabians, the intellectual elite of the British Labour party) often has radically different beliefs than those it tells its outer ring of supporters (in the Labour party). To rank-and-file workers, Fabians promise jobs that pay well enough for every worker to have the size family he chooses. But within the inner ring, a chillingly different attitude exists.

In an 1896 Fabian tract entitled "The Difficulties of Individualism," Sidney Webb looked at British society and came to a conclusion that, in its second part, is startlingly similar to that Chesterton ascribed to the viler sort of capitalist. Private property, Webb wrote, promotes "wrong production, both of commodities and of human beings; the preparation of senseless luxuries whilst there is need for more bread; and the breeding of degenerate hordes of demoralized "residuum" unfit for social life." Later in the tract he noted that one result of poverty is "a horde of semi-barbarians," and still later he warned of the "indiscriminate multiplication of the unfit." Socialism, Webb told the Fabian inner ring, must take property from the idle rich *and* children from the unfit poor.

Needless to say, if they wanted to ride into power on the vote of that "horde of semi-barbarians," the Fabians had to trumpet the first agenda far more loudly than the second. As a result the Fabians and their ideological bedfellows, already proud of their go-slow approach to socialism, also took a go-slow attitude toward eugenics. It is in that context that you need to read Chesterton's argument that Wells was "the Eugenist who destroyed Eugenics." What Wells was objecting to was a backlash that might follow from the hysterical alarm about a rapidly multiplying unfit being used to push through eugenic legislation. As you may notice in the appendices, the cleverer eugenists (of whatever political persuasion) hoped to make parenthood more burdensome for ordinary working people and then use that frustration to pass still harsher measures directed at the poor. It's an agenda that remains with us to this day.

Another agenda also needs explaining. The birth control movement became important after World War I and, since much of Chesterton's book was planned or written before the war, he had little to say about it. By the time his book came out, however, it was on well on its way to becoming a major political and social force. As you will see in the appendices, birth controllers were quite similar to the eugenists with one major difference. The eugenists of that day were divided over the value of birth control. It was being adopted by the very segments of society they considered superior, lowering birthrates and having a harmful eugenic effect. Some eugenists feared that a further spread of birth control would do even more harm, rightly pointing out that the methods of birth control used then (primarily barrier techniques) would be less effective among the poor. Others took the opposite approach, arguing that, since it was impossible to keep birth control from the 'fit,' it had to be taken to the 'unfit.'

Birth controllers aligned themselves with the latter, though they had amiable relations with the former since all agreed on the 'menace' of the feeble-minded (who were the ancestors of many of us).

Most present-day historians paint eugenists as the bad guys and, when forced to bring in the link to birth controllers, portray the latter as well-intentioned women seduced by the powerful connections of eugenists. As you'll see from the appendices, the opposite was true. Whatever their opinion of birth control, eugenists—a predominately male group—agreed that encouraging the fit to have more children (positive eugenics) was as much a part of their agenda as coercing a lower birthrate from the unfit (negative eugenics). Bolstered by positive measures, they didn't need to be as ruthless with negative measures as birth controllers, hence the latter's viciousness. Those who doubt that need only compare the first six chapters (by eugenists) with the last three.

If he had written a bit later, Chesterton would have delighted to take on birth controllers. The movement was just the sort of foe he loved. It was a bizarre blend of feminists with sexual axes to grind and society women obsessed with maintaining their privileged social positions. A common phobia about the 'breeding' going on in crowded slums united sexual radical with great wealth (and still does).

Birth controllers had their differences. Some were willing to tolerate 'soft' encouragements that make it easier for *their* sort of woman to have an extra child or two. (Subsidized child care fits that category.) In some remarks, Marie Stopes takes that approach. Others, particularly Margaret Sanger (in America), rejected positive eugenics outright, labeling it a "cradle competition" between fit and unfit. The only eugenics these women would tolerate were negative measures, and it was for that reason that they set up their clinics in poor neighborhoods. For ordinary women who want to be mothers and raise families, their worst enemies are often the very women who claim most loudly to speak for all women.

When present-day feminists talk darkly of "forced motherhood," they are, consciously or unconsciously, referring back to this long-hidden debate. The Darwinian worldview in which life is a struggle between fit and unfit is an iron-clad dogma among the left/liberal/progressive allies of feminism. If the birthrate of inferiors can't be forced down, then their male allies are likely to return to their old demand for "more children from the fit." It is that fear which drives feminists—some so unsuited for motherhood no sane person would force them into it—to fear any change in the measures put into place to reduce the birthrates of socially troublesome groups (particularly legalized abortion). That's also why feminists show so little interest in allowing less-affluent women to make other parenting choices—such as the type of school, public or private, secular or religious—that their child attends. Poor schools, crime-ridden neighborhoods, and irresponsible males are all forces that drive disadvantaged women into abortion clinics, just as they once did birth control clinics

Chesterton got in right in the second chapter when he described most eugenists as foolish people unable to understand the forces they were unleashing, forces that would be exploited by others far more ruthless than they. If you read enough eugenic literature, you'll soon detect something almost pitiful about the world in which they lived. In all too many cases, two or three generations back there was Someone Important about whom they were inordinately proud. But the family blood had grown thin since, and they must cling to that fading past and, with it, their claims to superior genes. Sensing this deep-seated inadequacy, birth controllers often gloated and pointed to how few (if any) descendants these pitiable men had.

History, as written by historians, has been lopsided in its condemnation of eugenists. The playful ridicule Chesterton heaps on them is much more accurate. Of all the early twentieth century groups afflicted with the sorts of scientific bigotry so easily erected on the ideas of Malthus and Darwin, they were the most benign. By claiming to be women helping other women, the agenda of birth controllers was far more deceptive. By rendering men and woman incapable of ever becoming parents, those who advocated forced sterilization dehumanized their victim far more. And in hindsight, the worst of all were groups such as the Immigration Restriction League, which was heavy with Harvard blue bloods. (See John Higham's *Strangers in the Land.*) For it was they who trapped millions of Jews in Eastern Europe where they faced death in Hitler's camps. Those who think this connection overdrawn should read a speech that Senator Henry Cabot Lodge gave to Congress on March 16, 1896 in support of immigration restriction. Fully 70 percent of his speech praised the superiority of Europe's "Germanic tribes." In reference to Jews and Slavs, Lodge warned, "If a lower race mixes with a higher in sufficient numbers, history teaches us that the lower will prevail." Neither Hitler nor our home-grown Ku Klux Klan could have said it better.

If the eugenists were the most benign of the movements intending to rebuild the world according the Malthus and Darwin, why have they drawn the most criticism? Widespread revulsion over the horrors of Nazism meant that one group had to take the blame. As the weakest and least useful, eugenists were an obvious pick. In vague and euphemistic prose, eugenists taught those who thought in Darwinian terms what the social and political implications of evolution were. Once that task was complete, their role was complete. Those they taught would carry the agenda far further than eugenics's early supporters had intended.

Keep in mind, however, that this secondary role doesn't excuse what the eugenists did. After all, it's not that eugenists were good and their fellow travelers were evil. It's that most eugenists were the sort who spent their lives talking rather than doing. Their muddled thought was a result of not having to keep a business alive in the midst of fierce competition (capitalists), win bitterly contested elections (liberals and socialists), or fight peer pressures that would force them to be mothers (feminists). The others were fiercer because they lived in a nastier world.

Of course, not all eugenists were harmless. Some, such as the more recent Dr. Alan Guttmacher, did quite a bit to further the collective agenda. But the great achievements in which he paid a major role (federally funded population control and legalized abortion) came while he was President of Planned Parenthood–World Population (1962–74) and not while he was Vice-President of the American Eugenics Association. When he took over the organization that Margaret Sanger founded, his eugenic agenda could easily be concealed behind rhetoric about helping women. He need not fear that the mass media would expose his eugenic past and was assured of the generous support of giant foundations, as well as a vast cadre of those—mostly on political left—who view people as things to be manipulated. A hint at just how dangerous Dr. Guttmacher was appeared on page 11 in the June 6, 1969 issue of *Medical World News* (a news magazine for physicians). There Guttmacher warned: "Each country will have to decide its own form of coercion, and determine how it is to be employed. At present the means available are compulsory sterilization and compulsory abortion. Perhaps some day a way of enforcing compulsory birth control will be feasible." That "some day" came in the early 1990s when Depo-Provera, a long-term, injectable contraceptive, became widely available.

With those all too contemporary remarks, I leave readers to enjoy Chesterton's most stimulating book.

Michael W. Perry
Seattle, October 9, 2000

# To the Reader

I publish these essays at the present time for a particular reason connected with the present situation; a reason which I should like briefly to emphasise and make clear.

Though most of the conclusions, especially towards the end, are conceived with reference to recent events, the actual bulk of preliminary notes about the science of Eugenics were written before the war. It was a time when this theme was the topic of the hour; when eugenic babies (not visibly very distinguishable from other babies) sprawled all over the illustrated papers; when the evolutionary fancy of Nietzsche was the new cry among the intellectuals; and when Mr. Bernard Shaw and others were considering the idea that to breed a man like a cart-horse was the true way to attain that higher civilisation of intellectual magnanimity and sympathetic insight which may be found in cart-horses. It may therefore appear that I took the opinion too controversially, and it seems to me that I sometimes took it too seriously. But the criticism of Eugenics soon expanded of itself into a more general criticism of a modern craze for scientific officialism and strict social organisation.

And then the hour came when I felt, not without relief, that I might well fling all my notes into the fire. The fire was a very big one, and was burning up bigger things than such pedantic quackeries. And, anyhow, the issue itself was being settled in a very different style. Scientific officialism and organisation in the State which had specialised in them, had gone to war with the older culture of Christendom. Either Prussianism would win and the protest would be hopeless, or Prussianism would lose and the protest would be needless. As the war advanced from poison gas to piracy against neutrals, it grew more and more plain that the scientifically organised State was not increasing in popularity. Whatever happened, no Englishmen would ever again go nosing round the stinks of that low laboratory, So I thought all I had written irrelevant, and put it out of my mind.

I am greatly grieved to say that it is not irrelevant. It has gradually grown apparent, to my astounded gaze, that the ruling classes in England are still proceeding on the assumption that Prussia is a pattern for the whole world. If parts of my book are nearly nine years old, most of their principles and proceedings are a great deal older. They can offer us nothing but the same stuffy science, the same bullying bureaucracy and the same terrorism by tenth-rate professors that have led the German Empire to its recent conspicuous triumph. For that reason, three years after the war with Prussia, I collect and publish these papers.

G. K. C.

# 1

# What Is Eugenics?

The point here is that a new school believes Eugenics *against* Ethics. And it is proved by one familiar fact: that the heroisms of history are actually the crimes of Eugenics.

The wisest thing in the world is to cry out before you are hurt. It is no good to cry out after you are hurt; especially after you are mortally hurt. People talk about the impatience of the populace; but sound historians know that most tyrannies have been possible because men moved too late. It is often essential to resist a tyranny before it exists. It is no answer to say, with a distant optimism, that the scheme is only in the air. A blow from a hatchet can only be parried while it is in the air.

### Eugenic Tyranny

But it is quite certain that no existing democratic government would go as far as we Eugenists think right in the direction of limiting the liberty of the subject for the sake of the racial qualities of future generations.

—LEONARD DARWIN, CAMBRIDGE UNIVERSITY EUGENICS SOCIETY, 1912

There exists to-day a scheme of action, a school thought, as collective and unmistakable as any of those by whose grouping alone we can make any outline of history. It is as firm a fact as the Oxford Movement, or the Puritans of the Long Parliament; or the Jansenists; or the Jesuits. It is a thing that can be pointed out; it is a thing that can be discussed; and it is a thing that can still be destroyed. It is called for convenience "Eugenics"; and that it ought to be destroyed I propose to prove in the pages that follow. I know that it means very different things to different people; but that is only because evil always takes advantage of ambiguity. I know it is praised with high professions of idealism and benevolence; with silver-tongued rhetoric about purer motherhood and a happier posterity. But that is only because evil is always flattered, as the Furies were called "The

Gracious Ones." I know that it numbers many disciples whose intentions are entirely innocent and humane; and who would be sincerely astonished at my describing it as I do. But that is only because evil always wins through the strength of its splendid dupes; and there has in all ages been a disastrous alliance between abnormal innocence and abnormal sin. Of these who are deceived I shall speak of course as we all do of such instruments; judging them by the good they think they are doing, and not by the evil which they really do. But Eugenics itself does exist for those who have sense enough to see that ideas exist; and Eugenics itself, in large quantities or small, coming quickly or coming slowly, urged from good motives or bad, applied to a thousand people or applied to three, Eugenics itself is a thing no more to be bargained about than poisoning.

### Eugenics Defined

Eugenics is the science which deals with all influences that improve the inborn qualities of a race; also with those that develop them to the utmost advantage.—FRANCIS GALTON, 1904

## The Essence of Eugenics

It is not really difficult to sum up the essence of Eugenics: though some of the Eugenists seem to be rather vague about it. The movement consists of two parts: a moral basis, which is common to all, and a scheme of social application which varies a good deal. For the moral basis, it is obvious that man's ethical responsibility varies with his knowledge of the consequences. If I were in charge of a baby (like Dr. Johnson in that tower of vision), and if the baby was ill through having eaten the soap, I might possibly send for a doctor, I might be calling him away from much more serious cases, from the bedsides of babies whose diet had been far more deadly; but I should be justified. I could not be expected to know enough about his other patients to be obliged (or even entitled) to sacrifice to them the baby for whom I was primarily and directly responsible. Now the Eugenic moral basis is this; that the baby for whom we are primarily and directly responsible is the babe unborn. That is, that we know (or may come to know) enough of certain tendencies in biology to consider the fruit of some contemplated union in that direct and clear light of conscience which we can now only fix on the other partner in that union. The one duty can conceivably be as definite as or more definite than the other. The baby that does not exist can be considered even before the wife who does. Now it is essential to grasp that this is a comparatively new note in morality. Of course sane people always thought the aim of marriage was the procreation of children to the glory of God or according to the plan of Nature; but whether they counted such children as God's reward for service or Nature's premium on sanity, they always left the reward to God or the premium to Nature, as a less definable thing. The only person (and this is the point) towards whom one could have precise duties was the partner in the process. Directly considering the partner's claims was the nearest one could get to indirectly considering the claims of posterity. If the women of the harem sang praises of the hero as the Moslem mounted his horse, it was because this was the due of a man; if the Christian knight helped his wife off her horse, it was because this was the due of a woman. Definite and detailed dues of this kind they did not predicate of the babe unborn; regarding him in that agnostic and opportunist light in which Mr. Browdie regarded the hypothetical child of Miss Squeers. Thinking these sex relations healthy, they naturally hoped they would produce healthy children; but that was all. The Moslem woman doubtless expected Allah to send beautiful sons to an obedient wife; but she would not have allowed any direct vision of such sons to alter the obedience itself. She would not have said, "I will now be a disobedient wife; as the learned leech informs me that great prophets are often the children of disobedient wives." The knight doubtless hoped that the saints would help him to strong children, if he did all the duties of his station, one of which might be helping his wife off her horse; but he would not have refrained from doing this because he had read in a book that a course of falling off horses often resulted in the birth of a genius. Both Moslem and Christian would have thought such speculations not only impious but utterly unpractical. I quite agree with them; but that is not the point here.

### Eugenic's Aim

The aim of eugenics is to bring as many influences as can be reasonably employed, to cause the useful classes in the community to contribute *more* than their proportion to the next generation.—FRANCIS GALTON, 1904

## Eugenics Against Ethics

The point here is that a new school believes Eugenics *against* Ethics. And it is proved by one familiar fact: that the heroisms of history are actually the crimes of Eugenics. The Eugenists' books and articles are full of suggestions that non-eugenic unions should and may come to be regarded as we regard sins; that we should really feel that marrying an invalid is a kind of cruelty to children. But history is full of the praises of people who have held sacred such ties to invalids; of cases like those of Colonel Hutchinson and Sir William Temple, who remained faithful to betrothals when beauty and health had been apparently blasted. And though the illnesses of Dorothy Osborne and Mrs. Hutchinson may not fall under the Eugenic speculations (I do not know), it is obvious that they might have done so; and certainly it would not have made any difference to men's moral opinion of the act. I do not discuss here which morality I favour; but I insist that they are opposite. The Eugenist really sets up as saints the very men whom hundreds of families have called sneaks. To be consistent, they ought to put up statues to the men who deserted their loves because of bodily misfortune; with inscriptions celebrating the good Eugenist who, on his fiancée falling off a bicycle, nobly refused to marry her; or to the young hero who, on hearing of an uncle with erysipelas, magnanimously broke his word. What is perfectly plain is this: that mankind have hitherto held the bond between man and woman so sacred, and the effect of it on the children so incalculable, that they have always admired the maintenance of honour more than the maintenance of safety. Doubtless they thought that even the children might be none the worse for not being the children of cowards and shirkers; but this was not the first thought, the first commandment. Briefly, we may say that while many moral systems have set restraints on sex almost as severe as any Eugenist could set, they have almost always had the character of securing the fidelity of the two sexes to each other, and leaving the rest to God. To introduce an ethic which makes that fidelity or infidelity vary with some calculation about heredity is that rarest of all things, a revolution that has not happened before.

### Banned Marriages

The passion of love seems so overpowering that it may be thought folly to try to direct its course. But plain facts do not confirm this view. Social influences of all kinds have immense power in the end, and they are very various. If unsuitable marriages from the eugenic point of view were banned socially, or even regarded with the unreasonable disfavour which some attach to cousin-marriages, very few would be made.—FRANCIS GALTON, 1904

It is only right to say here, though the matter should only be touched on, that many Eugenists would contradict this, in so far as to claim that there was a consciously Eugenic reason for the horror of those unions which begin with the celebrated denial to man of the privilege of marrying his grandmother. Dr. S. R. Steinmetz, with that creepy simplicity of mind with which the Eugenists chill the blood, remarks that "we do not yet know quite certainly" what were "the motives for the horror of" that horrible thing which is the agony of Oedipus. With entirely amiable intention, I ask Dr. S. R. Steinmetz to speak for himself. I know the motives for regarding a mother or sister as separate from other women; nor have I reached them by any curious researches. I found them where I found an analogous aversion to eating a baby for breakfast. I found them in a rooted detestation in soul to liking a thing in one way, when you already like it in another quite incompatible way. Now it is perfectly true that this aversion may have acted eugenically; and so had a certain ultimate confirmation and basis in the laws of procreation. But there really cannot be any Eugenist quite so dull as not to see that this is not a defence of Eugenics but a direct denial of Eugenics. If something which has been discovered at last by the lamp of learning is something which has been acted on first from the light of nature, this (so far as it goes) is not an argument for pestering people, but an argument for letting them alone. If men did not marry their grandmothers when it was, for all they knew, a most hygienic habit; if we know now that they instinctively avoided scientific peril; that, so far as it goes, is a point in favour of letting people marry anyone they like. It is simply the statement that sexual selection, or what Christians call falling in love, is a part of man which in the rough, and in long run can be trusted. And that is the destruction of the whole of this science at a blow.

### Eugenics Enlightened Future

Hence it is to be hoped that in the more enlightened future, a system will also be established for the examination of the family history of all those placed on the register as being unquestionably mentally abnormal, especially as regards the criminality, insanity, ill-health and pauperism of their relatives, and not omitting to note cases of marked ability. If all this were done it can hardly be doubted that many strains would be discovered which no one could deny ought to be made to die out in the interest of the nation; and in this way the necessity for legislation, such as that proposed by the Royal Commission on the care and control of the feeble-minded, would be further emphasised.

—LEONARD DARWIN, CAMBRIDGE UNIVERSITY EUGENICS SOCIETY, 1912

The second part of the definition, the persuasive or coercive methods to be employed, I shall deal with more fully in the second part of this book. But some such summary as the following may here be useful. Far into the unfathomable past of our race we find the assumption that the founding of a family is the personal adventure of a free man. Before slavery sank slowly out of sight under the new climate of Christianity, it may or may not be true that slaves were in some sense bred like cattle, valued as a promising stock for labour. If it was so it was so in a much looser and vaguer sense than the breeding of the Eugenists; and such modem philosophers read into the old paganism a fantastic pride and cruelty which are wholly modern. It may be, however, that pagan slaves had some shadow of the blessings of the Eugenist's care. It is quite certain that the pagan freemen would have killed the first man that suggested it. I mean suggested it seriously; for Plato was only a Bernard Shaw who unfortunately made his jokes in Greek. Among free men, the law, more often the creed, most commonly of all the custom, have laid all sorts of restrictions on sex for this reason or that. But law and creed and custom have never concentrated heavily except upon fixing and keeping the family when once it had been made. The act of founding the family, I repeat, was an individual adventure outside the frontiers of the State. Our first forgotten ancestors left this tradition behind them; and our own latest fathers and mothers a few years ago would have, thought us lunatics to be discussing it. The shortest general definition of Eugenics on its practical side is that it does, in a more or less degree, propose to control some families at least as if they were families of pagan slaves. I shall discuss later the question of the people to whom this pressure may be applied; and the much more puzzling question of what people will apply it. But it is to be applied at the very least by somebody to somebody, and that on certain calculations about breeding which are affirmed to be demonstrable. So much for the subject itself. I say that this thing exists. I define it as closely as matters involving moral evidence can be defined; I call it Eugenics. If after that anyone chooses to say that Eugenics is not the Greek for this—I am content to answer "chivalrous" "is not the French for "horsy"; that such controversial games are more horsy than chivalrous.

### Present Day People Like Stray Dogs

Men and women of the present day are, to those we might hope to bring into existence, what the pariah dogs of the streets of an Eastern town are to our own highly bred varieties.—FRANCIS GALTON, 1865

### Breeding Genius Like Horses and Cattle

If the twentieth part of the cost and pains were spent in measures for the improvement of the human race that is spent on the improvement of the breed of horses and cattle, what a galaxy of genius might we not create!
—FRANCIS GALTON, 1865

CHAPTER

2

# The First Obstacles

> I will call it the Feeble-Minded Bill, both for brevity and because the description is strictly accurate. It is, quite simply and literally, a Bill for incarcerating as madmen those whom no doctor will consent to call mad. It is enough if some doctor or other may happen to call them weak-minded.

Now before I set about arguing these things, there is a cloud of skirmishers, of harmless and confused modern sceptics, who ought to be cleared off or calmed down before we come to debate with the real doctors of the heresy. If I sum up my statement thus: "Eugenics, as discussed, evidently means the control of some men over the marriage and unmarriage of others; and probably means the control of the few over the marriage and unmarriage of the many," I shall first of all receive the sort of answers that float like skim on the surface of teacups and talk. I may very roughly and rapidly divide these preliminary objectors into five sects; whom I will call the Euphemists, the Casuists, the Autocrats, the Precedenters, and the Endeavourers. When we have answered the immediate protestation of all these good, shouting, short-sighted people, we can begin to do justice to those intelligences that are really behind the idea.

## Eugenists Who Are Euphemists
Most Eugenists are Euphemists. I mean merely that short words startle them, while long words soothe them. And they are utterly incapable of translating the one into the other, however obviously they mean the same thing. Say to them "The persuasive and even coercive powers of the citizen should enable him to make sure that the burden of longevity in the previous generation does not become disproportionate and intolerable, especially to the females"; say this to them and they will sway slightly to and fro like babies sent to sleep in cradles. Say to them "Murder your

mother," and they sit up quite suddenly. Yet the two sentences, in cold logic, are exactly the same. Say to them "It is not improbable that a period may arrive when the narrow if once useful distinction between the anthropoid homo and the other animals, which has been modified on so many moral points, may be modified also even in regard to the important question of the extension of human diet"; say this to them, and beauty born of murmuring sound will pass into their face. But say to them, in a simple, manly, hearty way "Let's eat a man!" and their surprise is quite surprising. Yet the sentences say just the same thing. Now, if anyone thinks these two instances extravagant, I will refer to two actual cases from the Eugenic discussions. When Sir Oliver Lodge spoke of the methods "of the studfarm" many Eugenists exclaimed against the crudity of the suggestion. Yet long before that one of the ablest champions in the other interest had written "What nonsense this education is! Who could educate a racehorse or a greyhound?" Which most certainly either means nothing, or the human studfarm. Or again, when I spoke of people "being married forcibly by the police," another distinguished Eugenist almost achieved high spirits in his hearty assurance that no such thing had ever come into their heads, Yet a few days after I saw a Eugenist Pronouncement, to the effect that the State ought extend its Powers in this area. The State can only be that corporation which men permit to employ compulsion; and this area can only be the area of sexual selection. I mean somewhat more than an idle jest when I say that the policeman will generally be found in that area. But I willingly admit that the policeman who looks after weddings will be like the policeman who looks after wedding-presents. He will be in plain clothes. I do not mean that a man in blue with a helmet will drag the bride and bridegroom to the altar. I do mean that nobody that man in blue is told to arrest will even dare to come near the church. Sir Oliver did not mean that men would be tied up in stables and scrubbed down by grooms. He meant that they would undergo a loss of liberty which to men is even more infamous. He meant that the only formula important to Eugenists would be "by Smith out of Jones." Such a formula is one of the shortest in the world; and is certainly the shortest way with the Euphemists.

### People as Plastic as Clay

The power of man over animal life, in producing whatever varieties of form he pleases, is enormously great. It would seem as though the physical structure of future generations was almost as plastic as clay, under the control of the breeder's will. It is my desire to show, more pointedly than—so far as I am aware—has been attempted before, that [human] mental qualities are equally under control.—FRANCIS GALTON, 1865

*EUGENICS AND OTHER EVILS*

## Eugenists Who Are Casuists

The next sect of superficial objectors is even more irritating. I have called them, for immediate purposes, the Casuists. Suppose I say "I dislike this spread of Cannibalism in the West End restaurants." Somebody is sure to say "Well, after all, Queen Eleanor when she sucked blood from her husband's arm was a cannibal." What is one to say to such people? One can only say "Confine yourself to sucking poisoned blood from people's arms, and I permit you to call yourself by the glorious title of Cannibal." In this sense people say of Eugenics, "After all, whenever we discourage a schoolboy from marrying a mad negress with a hump back, we are really Eugenists." Again one can only answer, "Confine yourselves strictly to such schoolboys as are naturally attracted to hump-backed negresses; and you may exult in the title of Eugenist, all the more proudly because that distinction will be rare." But surely anyone's common-sense must tell him that if Eugenics dealt only with such extravagant cases, it would be called common-sense—and not Eugenics. The human race has excluded such absurdities for unknown ages, and has never yet called it Eugenics. You may call it flogging when you hit a choking gentleman on the back; you may call it torture when a man unfreezes his fingers at the fire; but if you talk like that a little longer you will cease to live among living men. If nothing but this mad minimum of accident were involved, there would be no such thing as a Eugenic Congress, and certainly no such thing as this book.

### Marriages as Crimes

The Church and the State and public opinion may permit the marriage of the feeble-minded girl of sixteen, or a marriage between a diseased inebriate and a maiden clear-eyed like the dawn; but the eugenist has regard to the end thereof, and he is false to his creed if he does not declare that these are crimes and outrages perpetrated alike upon the living and the unborn.

—C.W. SALEEBY, 1914

## Eugenists Who Are Autocrats

I had thought of calling the next sort of superficial people the Idealists; but I think this implies a humility towards impersonal good they hardly show; so I call them the Autocrats. They are those who give us generally to understand that every modern reform will "work" all right, because they will be there to see. Where they will be, and for how long, they do not explain very clearly. I do not mind their looking forward to numberless lives in succession; for that is the shadow of a human or divine hope. But even a theosophist does not expect to be a vast number of people at once. And these people most certainly propose to be responsible for a whole movement after it has left their hands. Each man promises to be about a thousand policemen. If you ask them how this or that will work, they will answer, "Oh, I would certainly insist on this"; or "I would never go so far as that"; as if they could return to this earth and do what no ghost has ever done quite successfully—force men to forsake their sins. Of these it is enough to say that they do not understand the nature of a law any more than the nature of a dog. If you let loose a law, it will do as a dog does. It will obey its own nature, not yours. Such sense as you have put into the law (or the dog) will be fulfilled. But you will not be able to fulfil a fragment of anything you have forgotten to put into it.

Along with such idealists should go the strange people who seem to think that you can consecrate and purify any campaign for ever by repeating the names of the abstract virtues that its better advocates had in mind. These people will say "So far from aiming at *slavery*, the Eugenists are seeking *true* liberty; liberty from disease and degeneracy, etc." Or they will say "We can assure Mr. Chesterton that the Eugenists have *no* intention of segregating the harmless; justice and mercy are the very motto of—" etc. To this kind of thing perhaps the shortest answer is this. Many of those who speak thus are agnostic or generally unsympathetic to official religion. Suppose one of them said "The Church of England is full of hypocrisy." What would he think of me if I answered, "I assure you that hypocrisy is condemned by every form of Christianity; and is particularly repudiated in the Prayer Book"? Suppose he said that the Church of Rome had been guilty of great cruelties. What would he think of me if I answered, "The Church is expressly bound to meekness and charity; and therefore cannot be cruel"? This kind of people need not detain us long.

### The Abolition of Marriage

Thirdly, the people called eugenists do not seek the abolition of marriage. They indeed assert their intention of judging all human institutions by their supreme criterion—the quality of the human life they produce—and thus they may condemn certain aspects of marriage as we practise it. Undoubtedly the eugenist declines to accept conventional, legal, or ecclesiastical standards of judgment in this or any other matter, but inquiry compels him to recognise in marriage the foremost and most fundamental instrument of his purpose. . . . If this is to "attack marriage," then he does attack marriage. But this is rather to make a stand for marriage against the influences which now threaten to destroy it.—C.W. SALEEBY, 1914

### Eugenists As Precedenters

Then there are others whom I may call the Precedenters; who flourish particularly in Parliament. They are best represented by the solemn official who said the other day that he could not understand the clamour against the Feeble-Minded Bill, as it only extended the principles of the old Lunacy Laws. To which again one can only answer "Quite so. It only extends the principles of the Lunacy Laws to persons without a trace of lunacy." This lucid politician finds an old law, let us say, about keeping lepers in quarantine. He simply alters the word "lepers" to "long-nosed people," and says blandly that the principle is the same.

### Drawing a Line

The real practical question as to how to select the individuals who should be segregated—how actually to draw the line—has, it may truly be said, still been shirked in this discussion. This is no doubt true. But it must be remembered that in many similar cases in the practical affairs of life no rules indicating exactly where the line should be drawn can be laid down in words even where the most vital decisions have to be made. This is true, for instance, in many respects in deciding whether or not a man is a lunatic, an idiot, or a criminal. The answer given to such questions must in reality always depend on the judgment of men merely guided by the knowledge of broad and general principles; and in drawing the line with regard to the segregation of the Feeble-Minded or of any other class of the eugenically unfit in the interests of posterity without laying down any hard and fast rules, it can merely be said that such a proceeding would form no exception to the methods generally adopted in such cases. Under an all-wise government these guiding principles might perhaps be such as would result in the net to catch the mentally defective being spread very widely.—LEONARD DARWIN, *EUGENICS REVIEW*, 1912

## Eugenists Who Are Endeavourers

Perhaps the weakest of all are those helpless persons whom I have called the Endeavourers. The prize specimen of them was another M.P. who defended the same Bill as "an honest attempt" to deal with a great evil: as if one had a right to dragoon and enslave one's fellow citizens as a kind of chemical experiment; in a state of reverent agnosticism about what would come of it. But with this fatuous notion that one can deliberately establish the Inquisition or the Terror, and then faintly trust the larger hope, I shall have to deal more seriously in a subsequent chapter. It is enough to say here that the best thing the honest Endeavourer could do would be to make an honest attempt to know what he is doing. And not to do anything else until he has found out.

### Some Way or Other

I hence conclude that the improvement of the breed of mankind is no insuperable difficulty. If everybody were to agree on the improvement of the race of men being a matter of the very utmost importance, and if the theory of the hereditary transmission of qualities in men was as thoroughly understood as it

is in the case of our domestic animals, I see no absurdity in supposing that, in some way or other, the improvement would be carried into effect.
—FRANCIS GALTON, 1865

## Eugenists Who Defy Naming

Lastly, there is a class of controversialists so hopeless and futile that I have really failed to find a name for them. But whenever anyone attempts to argue rationally for or against any existent and recognisable *thing*, such as the Eugenic class of legislation, there are always people who begin to chop hay about Socialism and Individualism; and say "*You* object to all State interference; I am in favour of State interference. *You* are an Individualist; *I*, on the other hand," etc. To which I can only answer, with heart-broken patience, that I am not an Individualist, but a poor fallen but baptized journalist who is trying to write a book about Eugenists, several of whom he has met; whereas he never met an Individualist, and is by no means certain he would recognise him if he did. In short, I do not deny, but strongly affirm, the right of the State. to interfere to cure a great evil. I say that in this case it would interfere to create a great evil; and I am not going to be turned from the discussion of that direct issue to bottomless botherations about Socialism and Individualism, or the relative advantages of always turning to the right and always turning to the left.

### Extreme Individualists

For the rest the Bill has been rearranged and for the most part simplified. Less has been left to the regulations, and throughout one can trace evidence of an endeavour to conciliate the opposition of the extreme individualists who so effectively delayed the proceedings in Committee last year.
—*EUGENICS REVIEW*, 1913

## Eugenists Who Fail To See a Revolution

And for the rest, there is undoubtedly an enormous mass of sensible, rather thoughtless people, whose rooted sentiment it is that any deep change in our society must be in some way infinitely distant. They cannot believe that men in hats and coats like themselves can be preparing a revolution; all their Victorian philosophy has taught them that such transformations are always slow. Therefore, when I speak of Eugenic legislation, or the coming of the Eugenic State, they think of it as something like *The Time Machine* or *Looking Backward*: a thing that, good or bad, will have to fit itself to their great-great-great-grandchild, who may be very different and may like it; and who in any case is rather a distant relative. To all this I have, to begin with, a very short and simple answer. The Eugenic State has begun. The first of the Eugenic Laws has already been adopted by the Government of this country; and passed with the applause of both parties through the dominant House of Parliament. This first Eugenic Law clears the ground and may be said to proclaim negative Eugenics; but it cannot be defended, and nobody has attempted to defend it, except on the Eugenic theory. I will call it the Feeble-Minded Bill, both for brevity and because the description is strictly accurate. It is, quite simply and literally, a Bill for incarcerating as madmen those whom no doctor will consent to call mad. It is enough if some doctor or other may happen to call them weak-minded. Since there is scarcely any human being to whom this term has not been conversationally applied by his own friends and relatives on some occasion or other (unless his friends and relatives have been lamentably lacking in spirit), it can be clearly seen that this law, like the early Christian Church (to which, however, it presents points of dissimilarity), is a net drawing in of all kinds. It must not be supposed that we have a stricter definition incorporated in the Bill. Indeed, the first definition of "feeble-minded" in the Bill was much looser and vaguer than the phrase "feeble-minded" itself. It is a piece of yawning idiocy about "persons who though capable of earning their living under favourable circumstances" (as if anyone could earn his living if circumstances were directly unfavourable to his doing so), are nevertheless "incapable of managing their affairs with proper prudence"; which is exactly what all the world and his wife are saying about their neighbours all over this planet. But as an incapacity for any kind of thought is now regarded as statesmanship, there is nothing so very novel about such slovenly drafting. What is novel and what is vital is this: that the *defence* of this crazy Coercion Act is a Eugenic defence. It is not only openly said, it is eagerly urged, that the aim of the measure is to prevent any person whom these propagandists do not happen to think intelligent from having any wife or children. Every tramp who is sulky, every labourer who is shy, every rustic who is eccentric, can quite easily be brought under such conditions as were designed for homicidal maniacs. That is the situation, and that is the point. England has forgotten the Feudal State; it is in the last anarchy of the Industrial State; there is much in Mr. Belloc's theory that it is approaching the Servile State; it cannot at present get at the Distributive State; it has almost certainly missed the Socialist State. But we are already under the Eugenist State; and nothing remains to us but rebellion.

# 3

# The Anarchy from Above

Anarchy is that condition of mind or methods in which you cannot stop
yourself. It is the loss of that self-control which can return to the normal.

A silent anarchy is eating out our society. I must pause upon the expression; because the true nature of anarchy is mostly misapprehended. It is
not in the least necessary that anarchy should be violent; nor is it necessary that it should come from below. A government may grow anarchic as
much as a people. The more sentimental sort of Tory uses the word anarchy as a mere term of abuse for rebellion; but he misses a most important
intellectual distinction. Rebellion may be wrong and disastrous; but even
when rebellion is wrong, it is never anarchy. When it is not self-defence,
it is usurpation. It aims at setting up a new rule in place of the old rule.
And while it cannot be anarchic in essence (because it has an aim), it certainly cannot be anarchic in method; for men must be organised when
they fight; and the discipline in a rebel army has to be as good as the discipline in the royal army. This deep principle of distinction must be
clearly kept in mind. Take for the sake of symbolism those two great spiritual stories which, whether we count them myths or mysteries, have so
long been the two hinges of all European morals. The Christian who is
inclined to sympathise generally with constituted authority will think of
rebellion under the image of Satan, the rebel against God. But Satan,
though a traitor, was not an anarchist. He claimed the crown of the cos-
mos; and had he prevailed, would have expected his rebel angels to give
up rebelling. On the other hand, the Christian whose sympathies are more
generally with just self-defence among the oppressed will think rather of
Christ Himself defying the High Priests and scourging the rich traders.
But whether or no Christ was (as some say) a Socialist, He most certainly
was not an Anarchist. Christ, like Satan, claimed the throne. He set up a
new authority against an old authority; but He set it up with positive com-

mandments and a comprehensible scheme. In this light all mediaeval people—indeed, all people until a little while ago—would have judged questions involving revolt. John Ball would have offered to pull down the government because it was a bad government, not because it was a government. Richard II would have blamed Bolingbroke not as a disturber of the peace, but as a usurper. Anarchy, then, in the useful sense of the word, is a thing utterly distinct from any rebellion, right or wrong. It is not necessarily angry; it is not, in its first stages, at least, even necessarily painful. And, as I said before, it is often entirely silent.

Anarchy is that condition of mind or methods in which you cannot stop yourself. It is the loss of that self-control which can return to the normal. It is not anarchy because men are permitted to begin uproar, extravagance, experiment, peril. It is anarchy when people cannot *end* these things. It is not anarchy in the home if the whole family sits up all night on New Year's Eve. It is anarchy in the home if members of the family sit up later and later for months afterwards. It was not anarchy in the Roman villa when, during the Saturnalia, the slaves turned masters or the masters slaves. It was (from the slave-owners' point of view) anarchy if, after the Saturnalia, the slaves continued to behave in a Saturnalian manner; but it is historically evident that they did not. It is not anarchy to have a picnic; but it is anarchy to lose all memory of meal times. It would, I think, be anarchy if (as is the disgusting suggestion of some) we all took what we liked off the sideboard. That is the way swine would eat if swine had sideboards; they have no immovable feasts; they are uncommonly progressive, are swine. It is this inability to return within rational limits after a legitimate extravagance that is the really dangerous disorder. The modern world is like Niagara. It is magnificent, but it is not strong. It is as weak as water—like Niagara. The objection to a cataract is not that it is deafening or dangerous or even destructive; it is that it cannot stop. Now it is plain that this sort of chaos can possess the powers that rule a society as easily as the society so ruled. And in modern England it is the powers that rule who are chiefly possessed by it—who are truly possessed by devils. The phrase, in its sound old psychological sense, is not too strong. The State has suddenly and quietly gone mad. It is talking nonsense; and it can't stop.

Now it is perfectly plain that government ought to have, and must have, the same sort of right to use exceptional methods occasionally that the private householder has to have a picnic or to sit up all night on New Year's Eve. The State, like the householder, is sane if it can treat such exceptions as exceptions. Such desperate remedies may not even be right; but such remedies are endurable as long as they are admittedly desperate.

Such cases, of course, are the communism of food in a besieged city; the official disavowal of an arrested spy; the subjection of a patch of civil life to martial law; the cutting of communication in a plague; or that deepest degradation of the commonwealth, the use of national soldiers not against foreign soldiers, but against their own brethren in revolt. Of these exceptions some are right and some wrong; but all are right in so far as they are taken as exceptions. The modern world is insane, not so much because it admits the abnormal as because it cannot recover the normal.

We see this in the vague extension of punishments like imprisonment; often the very reformers who admit that prison is bad for people propose to reform them by a little more of it. We see it in panic legislation like that after the White Slave scare, when the torture of flogging was revived for all sorts of ill defined and vague and variegated types of men. Our fathers were never so mad, even when they were torturers. They stretched the man out on the rack. They did not stretch the rack out, as we are doing. When men went witch-burning they may have seen witches everywhere—because their minds were fixed on witchcraft. But they did not see things to burn everywhere, because their minds were unfixed. While tying some very unpopular witch to the stake, with the firm conviction that she was a spiritual tyranny and pestilence, they did not say to each other, "A little burning is what my Aunt Susan wants, to cure her of backbiting," or "Some of these faggots would do your Cousin James good, and teach him to play with poor girls' affections."

Now the name of all this is Anarchy. It not only does not know what it wants, but it does not even know what it hates. It multiplies excessively in the more American sort of English newspapers. When this new sort of New Englander burns a witch the whole prairie catches fire. These people have not the decision and detachment of the doctrinal ages. They cannot do a monstrous action and still see it is monstrous. Wherever they make a stride they make a rut. They cannot stop their own thoughts, though their thoughts are pouring into the pit.

A final instance, which can be sketched much more briefly, can be found in this general fact: that the definition of almost every crime has become more and more indefinite, and spreads like a flattening and thinning cloud over larger and larger landscapes. Cruelty to children, one would have thought, was a thing about as unmistakable, unusual and appalling as parricide. In its application it has come to cover almost every negligence that can occur in a needy household. The only distinction is, of course, that these negligences are punished in the poor, who generally can't help them, and not in the rich, who generally can. But that is not the point I am arguing just now. The point here is that a crime we all instinctively connect with Herod on the bloody night of Innocents has come precious near being attributable to Mary and Joseph when they lost their child in the Temple. In the light of a fairly recent case (the confessedly kind mother who was lately jailed because her confessedly healthy children had no water to wash in) no one, I think, will call this an illegitimate literary exaggeration. Now this is exactly as if all the horror and heavy punishment, attached in the simplest tribes to parricide, could now be used against any son who had done any act that could colourably be supposed to have worried his father, and so affected his health. Few of us would be safe.

Another case out of hundreds is the loose extension of the idea of libel. Libel cases bear no more trace of the old and just anger against the man who bore false witness against his neighbour than "cruelty" do of the old and just horror of the parents that hated their own flesh. A libel case has become one of the sports of the less athletic rich—a variation of *baccarat*, a game of chance. A music-hall actress got damages for a song that was called "vulgar," which is as if I could fine or imprison my neighbour for calling my handwriting "rococo." A politician got huge damages because he was said to have spoken to children about Tariff Reform; as if that seductive topic would corrupt their virtue, like an indecent story. Sometimes libel is defined as anything calculated to hurt a man in his business; in which case any new tradesman calling himself a grocer slanders the grocer opposite. All this, I say, is Anarchy; for it is clear that its exponents possess no power of distinction, or sense of proportion, by which they can draw the line between calling a woman a popular singer and calling her a bad lot; or between charging a man with leading infants to Protection and leading them to sin and shame. But the vital point to which to return is this. That it is not necessarily, nor even specially, an anarchy in the populace. It is an anarchy in the organ of government. It is the magistrates—voices of the governing class—who cannot distinguish between cruelty and carelessness. It is the judges (and their very submissive special juries) who cannot see the difference between opinion and slander. And it is the highly placed and highly paid experts who have brought in the first Eugenic Law, the Feeble Minded Bill—thus showing that they can see no difference between a mad and a sane man.

### Quality Not Quantity

We must breed for quality, not quantity in the human species, or the end is nigh.—*BIRTH CONTROL NEWS*, AUG, 1922

That, to begin with, is the historic atmosphere in which this thing was born. It is a peculiar atmosphere, and luckily not likely to last. Real progress bears the same relation to it that a happy girl laughing bears to an hysterical girl who cannot stop laughing. But I have described this atmosphere first because it is the only atmosphere in which such a thing as the Eugenist legislation could be proposed among men. All other ages would have called it to some kind of logical account, however academic or narrow. The lowest sophist in the Greek schools would remember enough of Socrates to force the Eugenist to tell him (at least) whether Midias was segregated because he was curable or because he was incurable. The meanest Thomist of the mediaeval monasteries would have the sense to see that you cannot discuss a madman when you have not discussed a man. The most owlish Calvinist commentator in the seventeenth century would ask the Eugenist to reconcile such Bible texts as derided fools with the other Bible texts that praised them. The dullest shopkeeper in Paris in 1790 would have asked what were the Rights of Man, if they did not include the rights of the lover, the husband, and the father. It is only in our own London Particular (as Mr. Guppy said of the fog) that small figures can loom so large in the vapour, and even mingle with quite different figures, and have the appearance of a mob. But, above all, I have dwelt on the telescopic quality in these twilight avenues, because unless the reader realises how elastic and unlimited they are, he simply will not believe in the abominations we have to combat.

One of those wise old fairy tales, that come from nowhere and flourish everywhere, tells how a man came to own a small magic machine like a coffee-mill, which would grind anything he wanted when he said one word and stop when he said another. After performing marvels (which I wish my conscience would let me put into this book for padding) the mill was merely asked to grind a few grains of salt at an officers' mess on board ship; for salt is the type everywhere of small luxury and exaggeration, and sailors' tales should be taken with a grain of it. The man remembered the word that started the salt mill, and then, touching the word that stopped it, suddenly remembered that he forgot. The tall ship sank, laden and sparkling to the topmasts with salt like Arctic snows; but the mad mill was still grinding at the ocean bottom, where all the men lay drowned. And that (so says this fairy tale) is why the great waters about our world have a bitter taste. For the fairy tales knew what the modern mystics don't—that one should not let loose either the supernatural or the natural.

# 4

# The Lunatic and the Law

> Now this is the first weakness in the case of the Eugenists: that they cannot define who is to control whom; they cannot say by what authority they do these things.

The modern evil, we have said, greatly turns on this: that people do not see that the exception proves the rule. Thus it may or may not be right to kill a murderer; but it can only conceivably be right to kill a murderer because it is wrong to kill a man. If the hangman, having got his hand in, proceeded to hang friends and relatives to his taste and fancy, he would (intellectually) unhang the first man, though the first man might not think so. Or thus again, if you say an insane man is irresponsible, you imply that a sane man is responsible. He is responsible for the insane man. And the attempt of the Eugenists and other fatalists to treat all men as irresponsible is the largest and flattest folly in philosophy. The Eugenist has to treat everybody, including himself, as an exception to a rule that isn't there.

The Eugenists, as a first move, have extended the frontiers of the lunatic asylum: let us take this as our definite starting point, and ask ourselves what lunacy is, and what is its fundamental relation to human society. Now that raw juvenile scepticism that clogs all thought with catchwords may often be heard to remark that the mad are only the minority, the sane only the majority. There is a neat exactitude about such people's nonsense; they seem to miss the point by magic. The mad are not a minority because they are not a corporate body; and that is what their madness means. The sane are not a majority; they are mankind. And mankind (as its name would seem to imply) is a *kind,* not a degree. In so far as the lunatic differs, he differs from all minorities and majorities in kind.

The madman who thinks he is a knife cannot go into partnership with the other who thinks he is a fork. There is no trysting place outside reason; there is no inn on those wild roads that are beyond the world.

The madman is not he that defies the world. The saint, the criminal, the martyr, the cynic, the nihilist may all defy the world quite sanely. And even if such fanatics would destroy the world, the world owes them a strictly fair trial according to proof and public law. But the madman is not the man who defies the world; he is the man who *denies* it. Suppose we are all standing round a field and looking at a tree in the middle of it. It is perfectly true that we all see it (as the decadents say) in infinitely different aspects: that is not the point; the point is that we all say it is a tree. Suppose, if you will, that we are all poets, which seems improbable; so that each of us could turn his aspect into a vivid image distinct from a tree. Suppose one says it looks like a green cloud and another like a green fountain, and a third like a green dragon and the fourth like a green cheese. The fact remains: that they all say it looks like these things. It is a tree. Nor are any of the poets in the least mad because of any opinions they may form, however frenzied, about the functions or future of the tree. A conservative poet may wish to clip the tree; a revolutionary poet may wish to burn it. An optimist poet may want to make it a Christmas tree and hang candles on it. A pessimist poet may want to hang himself on it. None of these are mad, because they are all talking about the same thing. But there is another man who is talking horribly about something else. There is a monstrous exception to mankind. Why he is so we know not; a new theory says it is heredity; an older theory says it is devils. But in any case, the spirit of it is the spirit that denies, the spirit that really denies realities. This is the man who looks at the tree and does not say it looks like a lion, but says that it is a lamp-post.

I do not mean that all mad delusions are as concrete as this, though some are more concrete. Believing your own body is glass is a more daring denial of reality than believing a tree is a glass lamp at the top of a pole. But all true delusions have in them this unalterable assertion—that what is not is. The difference between us and the maniac is not about how things look or how things ought to look, but about what they self-evidently are. The lunatic does not say that he ought to be King; Perkin Warbeck might say that. He says he is King. The lunatic does not say he is as wise as Shakespeare; Bernard Shaw might say that. The lunatic says he is Shakespeare. The lunatic does not say he is divine in the same sense as Christ; Mr. R. J. Campbell would say that. The lunatic says he is Christ. In all cases the difference is a difference about what is there; not a difference touching what should be done about it.

For this reason, and for this alone, the lunatic is outside public law. This is the abysmal difference between him and the criminal. The criminal admits the facts, and therefore permits us to appeal to the facts. We can so arrange the facts around him that he may really understand that agreement is in his own interests. We can say to him, "Do not steal apples from this tree, or we will hang you on that tree." But if the man really thinks one tree is a lamp-post and the other tree a Trafalgar Square fountain, we simply cannot treat with him at all. It is obviously useless to say, "Do not steal apples from this lamp-post, or I will hang you on that fountain." If a man denies the facts, there is no answer but to lock him up. He cannot speak our language: not that varying verbal language which often misses fire even with us, but that enormous alphabet of sun and moon and green grass and blue sky in which alone we meet, and by which alone we can signal to each other. That unique man of genius, George Macdonald, described in one of his weird stories two systems of space co-incident; so that where I knew there was a piano standing in a drawing-room you knew there was a rose-bush growing in a garden. Something of this sort is in small or great affairs the matter with the madman. He cannot have a vote, because he is the citizen of another country. He is a foreigner. Nay, he is an invader and an enemy; for the city he lives in has been super-imposed on ours.

Now these two things are primarily to be noted in his case. First, that we can only condemn him to a *general* doom, because we only know his *general* nature. All criminals, who do particular things for particular reasons (things and reasons which, however criminal, are always comprehensible), have been more and more tried for such separate actions under separate and suitable laws ever since Europe began to become a civilisation—and until the rare and recent re-incursions of barbarism in such things as the Indeterminate Sentence. Of that I shall speak later; it is enough for this argument to point out the plain facts. It is the plain fact that every savage, every sultan, every outlawed baron, every brigand-chief has always used this instrument of the Indeterminate Sentence, which has been recently offered us as something highly scientific and humane. All these people, in short, being barbarians, have always kept their captives captive until they (the barbarians) chose to think the captives were in a fit frame of mind to come out. It is also the plain fact that all that has been called civilisation or progress, justice or liberty, for nearly three thousand years, has had the general direction of treating even the captive as a free man, in so far as some clear case of some defined crime had to be shown against him. All law has meant allowing the criminal, within some limits or other, to argue with the law: as job was allowed, or rather challenged, to argue with God. But the criminal is, among civilised men, tried by one law for one crime for a perfectly simple reason: that the motive of the crime, like the meaning of the law, is conceivable to the common intelligence. A man is punished specially as a burglar, and not generally as a bad man, because a man may be a burglar and in many other respects not be a bad man. The act of burglary is punishable because it is intelligible. But when acts are unintelligible, we can only refer them to a general untrustworthiness, and guard against them by a general restraint. If a man breaks into a house to get a piece of bread, we can appeal to his reason in various ways. We can hang him for housebreaking; or again (as has occurred to some daring thinkers) we can give him a piece of bread. But if he breaks in, let us say, to steal the parings of other people's finger nails, then we are in a difficulty: we cannot imagine what he is going to do with them, and therefore cannot easily imagine what we are going to do with him. If a villain comes in, in cloak and mask, and puts a little arsenic in the soup, we can collar him and say to him distinctly, "You are guilty of Murder; and I will now consult the code of tribal law, under which we live, to see if this practice is not forbidden." But if a man in the same cloak and mask is found at midnight putting a little soda-water in the soup, what can we say? Our charge necessarily becomes a more general one. We can only observe, with a moderation almost amounting to weakness, "You seem to be the sort of person who will do this sort of thing." And then we can lock him up. The principle of the indeterminate sentence is the creation of the indeterminate mind. It does apply to the incomprehensible creature, the lunatic. And it applies to nobody else.

The second thing to be noted is this: that it is only by the unanimity of sane men that we can condemn this man as utterly separate. If he says a tree is a lamp-post he is mad; but only because all other men say it is a tree. If some men thought it was a tree with a lamp on it, and others thought it was a lamp-post wreathed with branches and vegetation, then it would be a matter of opinion and degree; and he would not be mad, but merely extreme. Certainly he would not be mad if nobody but a botanist could see it was a tree. Certainly his enemies might be madder than he, if nobody but a lamplighter could see it was not a lamp-post. And similarly a man is not imbecile if only a Eugenist thinks so. The question then raised would not be his sanity, but the sanity of one botanist or one lamplighter or one Eugenist. That which can condemn the abnormally foolish is not the abnormally clever, which is obviously a matter in dispute. That which can condemn the abnormally foolish is the normally foolish. It is when he begins to say and do things that even stupid people do not say or do, that we have a right to treat him as the exception and not the rule. It is only because we none of us profess to be anything more than man that we have authority to treat him as something less.

### One-Third of U.S. Defective

According to a recent calculation, made in one of the bulletins of the Eugenics Record Office, about one-third of the population in the United States is thus capable of conveying mental deficiency, the "insane tendency," epilepsy, or some other defect.—C.W. SALEEBY, 1914

Now the first principle behind Eugenics becomes plain enough. It is the proposal that somebody or something should criticise men with the same superiority with which men criticise madmen. It might exercise this right with great moderation; but I am not here talking about the exercise, but about the right. Its *claim* certainly is to bring all human life under the Lunacy Laws.

Now this is the first weakness in the case of the Eugenists: that they cannot define who is to control whom; they cannot say by what authority they do these things. They cannot see the exception is different from the rule—even when it is misrule, even when it is an unruly rule. The sound sense in the old Lunacy Law was this: that you cannot deny that a man is a citizen until you are practically prepared to deny that he is a man. Men, and only men, can be the judges of whether he is a man. But any private club of prigs can be judges of whether he ought to be a citizen. When once we step down from that tall and splintered peak of pure insanity we step on to a tableland where one man is not so widely different from another. Outside the exception, what we find is the average. And the practical, legal shape of the quarrel is this: that unless the normal men have the right to expel the abnormal, what particular sort of abnormal men have the right to expel the normal men? If sanity is not good enough, what is there that is saner than sanity?

Without any grip of the notion of a rule and an exception, the general idea of judging people's heredity breaks down and is useless. For this reason: that if everything is the result of a doubtful heredity, the judgment itself is the result of a doubtful heredity also. Let it judge not that it be not judged. Eugenists, strange to say, have fathers and mothers like other people; and our opinion about their fathers and mothers is worth exactly as much as their opinions about ours. None of the parents were lunatics, and the rest is mere likes and dislikes. Suppose Dr. Saleeby had gone up to Byron and said, "My lord, I perceive you have a club-foot and inordinate passions: such are the hereditary results of a profligate soldier marrying a hot-tempered woman." The poet might logically reply (with characteristic lucidity and impropriety), "Sir, I perceive you have a confused mind and an unphilosophic theory about other people's love affairs. Such are the hereditary delusions bred by a Syrian doctor marrying a Quaker lady from York." Suppose Dr. Karl Pearson had said to Shelley, "From what I see of your temperament, you are running great risks in forming a connection with the daughter of a fanatic and eccentric like Godwin." Shelley would be employing the strict rationalism of the older and stronger free thinkers, if he answered, "From what I observe of your mind, you are rushing on destruction in marrying the great-niece of an old corpse of a courtier and dilettante like Samuel Rogers." It is only opinion for opinion. Nobody can pretend that either Mary Godwin or Samuel Rogers was mad; and the general view a man may hold about the healthiness of inheriting their blood or type is simply the same sort of general view by which men do marry for love or liking. There is no reason to suppose that Dr. Karl Pearson is any better judge of a bridegroom than the bridegroom is of a bride.

### No Right to be a Parent

We [eugenists] can distinguish between the right to live and the right to become a Parent. . . . We can do our best for the life that is, but can follow Nature and transcend her by mercifully forbidding it to reproduce its defect.
—C.W. SALEEBY, 1914

An objection may be anticipated here, but it is very easily answered. It may be said that we do, in fact, call in medical specialists to settle whether a man is mad; and that these specialists go by technical and even secret tests that cannot be known to the mass of men. It is obvious that this is true; it is equally obvious that it does not affect our argument. When we ask the doctor whether our grandfather is going mad, we still mean mad by our own common human definition. We mean, is he going to be a certain sort of person whom all men recognise when once he exists. That certain specialists can detect the approach of him, before he exists, does not alter the fact that it is of the practical and popular madman that we are talking, and of him alone. The doctor merely sees a certain fact potentially in the future, while we, with less information, can only see it in the present; but his fact is our fact and everybody's fact, or we should not bother about it at all. Here is no question of the doctor bringing an entirely new sort of person under coercion, as in the Feeble-Minded Bill. The doctor can say, "Tobacco is death to you," because the dislike of death can be taken for granted, being a highly democratic institution; and it is the same with the dislike of the indubitable exception called madness. The doctor can say, "Jones has that twitch in the nerves, and he may burn down the house." But it is not the medical detail we fear, but the moral upshot. We should say, "Let him twitch, as long as he doesn't burn down the house." The doctor may say, "He has that look in the eyes, and he may take the hatchet and brain you all." But we do not object to the look in the eyes as such; we object to consequences which, once come, we should all call insane if there were no doctors in the world. We should say, "Let him look how he likes; as long as he does not look for the hatchet."

Now, that specialists are valuable for this particular and practical purpose, of predicting the approach of enormous and admitted human calamities, nobody but a fool would deny. But that does not bring us one inch nearer to allowing them the right to define what is a calamity; or to call things calamities which common sense does not call calamities. We call in the doctor to save us from death; and, death being admittedly an evil, he has the right to administer the queerest and most recondite pill which he may think is a cure for all such menaces of death. He has not the right to administer death, as the cure for all human ills. And as he has no moral authority to enforce a new conception of happiness, so he has no moral authority to enforce a new conception of sanity. He may know I am going mad; for madness is an isolated thing like leprosy; and I know nothing about leprosy. But if he merely thinks my mind is weak, I may happen to think the same of his. I often do.

In short, unless pilots are to be permitted to ram ships on to the rocks and then say that heaven is the only true harbour; unless judges are to be allowed to let murderers loose, and explain afterwards that the murder had done good on the whole; unless soldiers are to be allowed to lose battles and then point out that true glory is to be found in the valley of humiliation; unless cashiers are to rob a bank in order to give it an advertisement; or dentists to torture people to give them a contrast to their comforts; unless we are prepared to let loose all these private fancies against the public and accepted meaning of life or safety or prosperity or pleasure—then it is as plain as Punch's nose that no scientific man must be allowed to meddle with the public definition of madness. We call him in to tell us where it is or when it is. We could not do so, if we had not ourselves settled what it is.

As I wish to confine myself in this chapter to the primary point of the plain existence of sanity and insanity, I will not be led along any of the attractive paths that open here. I shall endeavour to deal with them in the next chapter. Here I confine myself to a sort of summary. Suppose a man's throat has been cut, quite swiftly and suddenly, with a table knife, at a small table where we sit. The whole of civil law rests on the supposition that we are witnesses; that we saw it; and if we do not know about it, who does? Now suppose all the witnesses fall into a quarrel about degrees of eyesight. Suppose one says be had brought his reading-glasses instead of his usual glasses; and therefore did not see the man fall across the table and cover it with blood. Suppose another says he could not be certain it was blood, because a slight colour-blindness was hereditary in his family. Suppose a third says he cannot swear to the uplifted knife, because his oculist tells him he is astigmatic, and vertical lines do not affect him as do horizontal lines. Suppose another says that dots have often danced before his eyes in very fantastic combinations, many of which were very like one gentleman cutting another gentleman's throat at dinner. All these things refer to real experiences. There is such a thing as myopia; there is such a thing as colour-blindness; there is such a thing as astigmatism; there is such a thing as shifting shapes swimming before the eyes. But what should we think of a whole dinner party that could give nothing except these highly scientific explanations when found in company with a corpse? I imagine there are only two things we could think: either that they were all drunk, or they were all murderers.

And yet there is an exception. If there were one man at table who was admittedly *blind,* should we not give him the benefit of the doubt? Should we not honestly feel that he was the exception that proved the rule? The very fact that he could not have seen would remind us that the other men must have seen. the very fact that he had no eyes must remind us of eyes. A man can be blind; a man can be dead; a man can be mad. But the comparison is necessarily weak, after all. For it is the essence of madness to be unlike anything else in the world: which is perhaps why so many men wiser than we have traced it to another.

Lastly, the literal maniac is different from all other persons in dispute in this vital respect: that he is the only person whom we can, with a final lucidity, declare that we do not want. He is almost always miserable himself, and he always makes others miserable. But this is not so with the mere invalid. The Eugenists would probably answer all my examples by taking the case of marrying into a family with consumption (or some other disease which they are fairly sure is hereditary) and asking whether such cases at least are not clear cases for Eugenic intervention. Permit me to point out to them that they once more make a confusion of thought. The sickness or soundness of a consumptive may be a clear and calculable matter. The happiness or unhappiness of a consumptive is quite another matter, and is not calculable at all. What is the good of telling people that if they marry for love, they may be punished by being the parents of Keats or the parents of Stevenson? Keats died young; but he had more pleasure in a minute than a Eugenist gets in a month. Stevenson had lung-trouble; and it may, for all I know, have been perceptible to the Eugenic eye even a generation before. But who would perform that illegal operation: the stopping of Stevenson? Intercepting a letter bursting with good news, confiscating a hamper full of presents and prizes, pouring torrents of intoxicating wine into the sea, all this is a faint approximation of the Eugenic inaction of the ancestors of Stevenson. This, however, is not the essential point; with Stevenson it is not merely a case of the pleasure we get, but of the pleasure he got. If he had died without writing a line, he would have had more red-hot joy than is given to most men. Shall I say of him, to whom I owe so much, let the day perish wherein he was born? Shall I pray that the stars of the twilight thereof be dark and it be not numbered among the days of the year, because it shut not up the doors of his mother's womb? I respectfully decline; like Job, I will put my hand upon my mouth.

### Southern Italians as a Low Grade Race

At the same time in the States, certain low-grade races such, for instance, as the Southern Italians, have an extremely high birth-rate, which, the lecturer maintained, formed too high a proportion of the American population.
—BIRTH CONTROL NEWS, JULY, 1922

### Sterilization in the U.S.

May I point out that the sterilization by law, although perhaps a novel ideal to the insular Briton, has been in existence in the other great English-speaking nation for a long time? Fifteen States in the U.S.A. enacted sterilization laws before the year 1920. The knowledge that others have taken this important national step may, perhaps, make it easier for English men and women to consider the subject freed from that shrinking fear induced by anything that is too novel.—DR. MARIE STOPES, BIRTH CONTROL NEWS, NOV 1922

### Sterilization in Chicago Courts

From Chicago we learn that "Sterilization of men and women who may be the parents of 'socially inadequate' children, as determined by experts, is advocated in a volume issued by the psychopathic laboratory of the Municipal Court of Chicago."

A model law to accomplish this, which Chief Justice Olson announced, will be presented to the Illinois General Assembly.

The "Socially Inadequate Classes" are defined as:—

(1) Feeble-minded; (2) insane (including the psychopathic; (3) criminalistic (including the delinquent and wayward); (4) epileptic; (5) inebriate (including drug habitues).—BIRTH CONTROL NEWS, FEB. 1923

5

# The Flying Authority

*Indeed one Eugenist, Mr. A. H. Huth, actually had a sense of humour, and admitted this. He thinks a great deal of good could be done with a surgical knife, if we would only turn him loose with one.*

It happened one day that an atheist and a man were standing together on a doorstep; and the atheist said, "It is raining." To which the man replied, "What is raining?": which question was the beginning of a violent quarrel and a lasting friendship. I will not touch upon any heads of the dispute, which doubtless included Jupiter Pluvius, the Neuter Gender, Pantheism, Noah's Ark, Mackintoshes, and the Passive Mood; but I will record the one point upon which the two persons emerged in some agreement. It was that there is such a thing as an atheistic literary style; that materialism may appear in the mere diction of a man, though he be speaking of clocks or cats or anything quite remote from theology. The mark of the atheistic style is that it instinctively chooses the word which suggests that things are dead things; that things have no souls. Thus they will not speak of waging war, which means willing it; they speak of the "outbreak of war," as if all the guns blew up without the men touching them. Thus those Socialists that are atheist will not call their international sympathy, sympathy; they will call it "solidarity," as if the poor men of France and Germany were physically stuck together like dates in a grocer's shop. The same Marxian Socialists are accused of cursing the Capitalists inordinately; but the truth is that they let the Capitalists off much too easily. For instead of saying that employers pay less wages, which might pin the employers to some moral responsibility, they insist on talking about the "rise and fall" of wages; as if a vast silver sea of sixpences and shillings was always going up and down automatically like the real sea at Margate. Thus they will not speak of reform, but of development; and they spoil their one honest and virile phrase, "the class war," by talking of it as no one in his wits can talk of a war, predicting its finish and final result as one calculates the coming of Christmas Day or the taxes. Thus, lastly (as we shall see touching our special subject-matter here) the atheist style in letters always avoids talking of love or lust, which are things alive, and calls marriage or concubinage "the relations of the sexes"; as if a man and a woman were two wooden objects standing in a certain angle and attitude to each other, like a table and a chair.

## Eugenic Grammar

Now the same anarchic mystery that clings round the phrase, "*il pleut*," clings round the phrase, "*il faut*." In English it is generally represented by the passive mood in grammar, and the Eugenists and their like deal especially in it; they are as passive in their statements as they are active in their experiments. Their sentences always enter tail first, and have no subject, like animals without heads. It is never "the doctor should cut off this leg" or "the policeman should collar that man." It is always "Such limbs should be amputated," or "Such men should be under restraint." Hamlet said, "I should have fatted all the region kites with this slave's offal." The Eugenist would say, "The region kites should, if possible, be fattened; and the offal of this slave is available for the dietetic experiment." Lady Macbeth said, "Give me the daggers; I'll let his bowels out." The Eugenist would say, "In such cases the bowels should, etc." Do not blame me for the repulsiveness of the comparisons. I have searched English literature for the most decent parallels to Eugenist language.

The formless god that broods over the East is called "Om." The formless god who has begun to brood over the West is called "On." But here we must make a distinction. The impersonal word *on* is French, and the French have a right to use it, because they are a democracy. And when a Frenchman says "one" he does not mean himself, but the normal citizen. He does not mean merely "one," but one and all. "*On n'a que sa Parole*" does not mean "*Noblesse oblige*," or "I am the Duke of Billingsgate and must keep my word." It means: "One has a sense of honour as one has a backbone: every man, rich or poor, should feel honourable": and this, whether possible or no, is the purest ambition of the republic. But when the Eugenists say, "Conditions must be altered" or "Ancestry should be investigated," or what not, it seems clear that they do not mean that the democracy must do it, whatever else they may mean. They do not mean that any man not evidently mad may be trusted with these tests and re-arrangements, as the French democratic system trusts such a man with a vote or a farm or the control of a family. That would mean that Jones and Brown, being both ordinary men, would set about arranging each other's marriages. And this state of affairs would seem a little elaborate, and it might occur even to the Eugenic mind that if Jones and Brown are quite capable of arranging each other's marriages, it is just possible that they might be capable of arranging their own.

This dilemma, which applies in so simple a case, applies equally to any wide and sweeping system of Eugenist voting; for though it is true that the community can judge more dispassionately than a man can judge in his own case, this particular question of the choice of a wife is so full of disputable shades in every conceivable case, that it is surely obvious that almost any democracy would simply vote the thing out of the sphere of voting, as they would any proposal of police interference in the choice of walking weather or of children's names. I should not like to be the politician who should propose a particular instance of Eugenics to be voted on by the French people. Democracy dismissed, it is here hardly needful to consider the other old models. Modern scientists will not say that George III, in his lucid intervals, should settle who is mad; or that the aristocracy that introduced gout shall supervise diet.

### Marriage without Parenthood

It is to parenthood on the part of the transmissibly unworthy that we object. Negative eugenics has no right to object to their living *or to their marrying.* This must be insisted upon. Hitherto marriage and parenthood have been regarded as synonymous or equivalent by writers on eugenics, and they have said that such and such persons must not marry, when what they meant was that these persons must not become parents.—C.W. SALEEBY, 1914

I hold it clear, therefore, if anything is clear about the business, that the Eugenists do not merely mean that the mass of common men should settle each other's marriages between them; the question remains, therefore, whom they do instinctively trust when they say that this or that ought to be done. What is this flying and evanescent authority that vanishes wherever we seek to fix it? Who is the man who is the lost subject that governs the Eugenist's verb? In a large number of cases I think we can simply say that the individual Eugenist means himself, and nobody else. Indeed one Eugenist, Mr. A. H. Huth, actually had a sense of humour, and admitted this. He thinks a great deal of good could be done with a surgical knife, if we would only turn him loose with one. And this may be true. A great deal of good could be done with a loaded revolver, in the hands of a judicious student of human nature. But it is imperative that the Eugenist should perceive that on that principle we can never get beyond a perfect balance of different sympathies and antipathies. I mean that I should differ from Dr. Saleeby or Dr. Karl Pearson not only in a vast majority of individual cases, but in a vast majority of cases in which they would be bound to admit that such a difference was natural and reasonable. The chief victim of these famous doctors would be a yet more famous doctor: that eminent though unpopular practitioner, Dr. Fell.

### Banning Marriage

Special provisions as to which it might be open to question whether they were primarily in the interest of the individual have been omitted, as for example clause 50 in the earlier Bill, which prohibited marriage with a defective. There is, however, an amendment proposed to reinstate this clause in Committee, but the matter has not yet been reached.—*EUGENICS REVIEW*, 1913

## The Danger of the Strong-Minded

To show that such rational and serious differences do exist, I will take one instance from that Bill which proposed to protect families and the public generally from the burden of feeble-minded persons. Now, even if I could share the Eugenic contempt for human rights, even if I could start gaily on the Eugenic campaign, I should not begin by removing feeble-minded persons. I have known as many families in as many classes as most men; and I cannot remember meeting any very monstrous human suffering arising out of the presence of such insufficient and negative types. There seem to be comparatively few of them; and those few by no means the worst burdens upon domestic happiness. I do not hear of them often; I do not hear of them doing much more harm than good; and in the few cases I know well they are not only regarded with human affection, but can be put to certain limited forms of human use. Even if I were a Eugenist, then I should not personally elect to waste my time locking up the feeble-minded. The people I should lock up would be the strong-minded. I have known hardly any cases of mere mental weakness making a family a failure; I have known eight or nine cases of violent and exaggerated force of character making a family a hell. If the strong-minded could be segregated it would quite certainly be better for their friends and families. And if there is really anything in heredity, it would be better for posterity too. For the kind of egoist I mean is a madman in a much more plausible sense than the mere harmless "deficient"; and to hand on the horrors of his anarchic and insatiable temperament is a much graver responsibility than to leave a mere inheritance of childishness. I would not arrest such tyrants, because I think that even moral tyranny in a few homes is better than a medical tyranny turning the state into a madhouse. I would not segregate them, because I respect a man's free-will and his front-door and his right to be tried by his peers. But since free-will is believed by Eugenists no more than by Calvinists, since front-doors are respected by Eugenists no more than by house-breakers, and since the Habeas Corpus is about as sacred to Eugenists as it would be to King John, why do not *they* bring light and peace into so many human homes by removing a demoniac from each of them? Why do not the promoters of the Feeble-Minded Bill call at the many grand houses in town or country where such nightmares notoriously are? Why do they not knock at the door and take the bad squire away? Why do they not ring the bell and remove the dipsomaniac prize-fighter? I do not know; and there is only one reason I can think of, which must remain a matter of speculation. When I was at school, the kind of boy who liked teasing halfwits was not the sort that stood up to bullies.

That, however it may be, does not concern my argument. I mention the case of the strong-minded variety of the monstrous merely to give one out of the hundred cases of the instant divergence of individual opinions the moment we begin to discuss who is fit or unfit to propagate. If Dr. Saleeby and I were setting out on a segregating trip together, we should separate at the very door; and if he had a thousand doctors with him, they would all go different ways. Everyone who has known as many kind and capable doctors as I have, knows that the ablest and sanest of them have a tendency to possess some little hobby or half-discovery of their own, as that oranges are bad for children, or that trees are dangerous in gardens, or that many more people ought to wear spectacles. It is asking too much of human nature to expect them not to cherish such scraps of originality in a hard, dull, and often heroic trade. But the inevitable result of it, as exercised by the individual Saleebys, would be that each man would have his favourite kind of idiot. Each doctor would be mad on his own madman. One would have his eye on devotional curates; another would wander about collecting obstreperous majors; a third would be the terror of animal-loving spinsters, who would flee with all their cats and dogs before him. Short of sheer literal anarchy, therefore, it seems plain that the Eugenist must find some authority other than his own implied personality. He must, once and for all, learn the lesson which is hardest for him and me and for all our fallen race—the fact that he is only himself.

### Medical Tyranny

We now pass from mere individual men who obviously cannot be trusted, even if they are individual medical men, with such despotism over their neighbours; and we come to consider whether the Eugenists have at all clearly traced any more imaginable public authority, any apparatus of great experts or great examinations to which such risks of tyranny could be trusted. They are not very precise about this either; indeed, the great difficulty I have throughout in considering what are the Eugenist's proposals is that they do not seem to know themselves. Some philosophic attitude which I cannot myself connect with human reason seems to make them actually proud of the dimness of their definitions and the uncompleteness of their plans. The Eugenic optimism seems to partake generally of the nature of that dazzled and confused confidence, so common in private theatricals, that it will be all right on the night. They have all the ancient despotism, but none of the ancient dogmatism. If they are ready to reproduce the secrecies and cruelties of the Inquisition, at least we cannot accuse them of offending us with any of that close and complicated thought, that and exact logic which narrowed the minds of the Middle Ages; they have discovered how to combine the hardening of the heart with a sympathetic softening of the head. Nevertheless, there is one large, though vague, idea of the Eugenists, which is an idea, and which we reach when we reach this problem of a more general supervision.

It was best presented perhaps by the distinguished doctor who wrote the article on these matters in that composite book which Mr. Wells edited, and called "The Great State." [Editor: This is a reference to Wells' 1912 *Socialism and the Great State*.] He said the doctor should no longer be a mere plasterer of paltry maladies, but should be, in his own words, "the health adviser of the community." The same can be expressed with even more point and simplicity in the proverb that prevention is better than cure. Commenting on this, I said that it amounted to treating all people who are well as if they were ill. This the writer admitted to be true, only adding that everyone is ill. To which I rejoin that if everyone is ill the health adviser is ill too, and therefore cannot know how to cure that minimum of illness. This is the fundamental fallacy in the whole business of preventive medicine. Prevention is not better than cure. Cutting off a man's head is not better than curing his headache; it is not even better than failing to cure it. And it is the same if a man is in revolt, even a morbid revolt. Taking the heart out of him by slavery is not better than leaving the heart in him, even if you leave it a broken heart. Prevention is not only not better than cure; prevention is even worse than disease. Prevention means being an invalid for life, with the extra exasperation of being quite well. I will ask God, but certainly not man, to prevent me in all my doings. But the decisive and discussable form of this is well summed tip in that phrase about the health adviser of society. I am sure that those who speak thus have something in their minds larger and more illuminating than the other two propositions we have considered. They do not mean that all citizens should decide, which would mean merely the present vague and dubious balance. They do not mean that all medical men should decide, which would mean a much more unbalanced balance.

They mean that a few men might be found who had a consistent scheme and vision of a healthy nation, as Napoleon had a consistent scheme and vision of an army. It is cold anarchy to say that all men are to meddle in all men's marriages. It is cold anarchy to say that any doctor may seize and segregate anyone he likes. But it is not anarchy to say that a few great hygienists might enclose or limit the life of all citizens, as nurses do with a family of children. It is not anarchy, it is tyranny; but tyranny is a workable thing. When we ask by what process such men could be certainly chosen, we are back again on the old dilemma of despotism, which means a man, or democracy which means men, or aristocracy which means favouritism. But as a vision the thing is plausible and even rational. It is rational, and it is wrong.

It is wrong, quite apart from the suggestion that an expert on health cannot be chosen. It is wrong because an expert on health cannot exist. An expert on disease can exist, for the very reason we have already considered in the case of madness, because experts can only arise out of exceptional things. A parallel with any of the other learned professions will make the point plain. If I am prosecuted for trespass, I will ask my solicitor which of the local lanes I am forbidden to walk in. But if my solicitor, having gained my case, were so elated that he insisted on settling what lanes I should walk in; if he asked me to let him map out all my country walks, because he was the perambulatory adviser of the community—then that solicitor would solicit in vain. If he will insist on walking behind me through woodland ways, pointing out with his walking-stick likely avenues and attractive short-cuts, I shall turn on him with passion, saying: "Sir, I pay you to know one particular puzzle in Latin and Norman-French, which they call the law of England; and you do know the law of England. I have never had any earthly reason to suppose that you know England. If you did, you would leave a man alone when he was looking at it." As are the limits of the lawyer's special knowledge about walking, so are the limits of the doctor's. If I fall over the stump of a tree and break my leg, as is likely enough, I shall say to the lawyer, "Please go and fetch the doctor." I shall do it because the doctor really has a larger knowledge of a narrower area. There are only a certain number of ways in which a leg can be broken; I know none of them, and he knows all of them. There is such a thing as being a specialist in broken legs. There is no such thing as being a specialist in legs. When unbroken, legs are a matter of taste. If the doctor has really mended my leg, he may merit a colossal equestrian statue on the top of an eternal tower of brass. But if the doctor has really mended my leg, he has no more rights over it. He must not come and teach me how to walk; because he and I learnt that in the same school, the nursery. And there is no more abstract likelihood of the doctor walking more elegantly than I do than there is of the barber or the bishop or the burglar walking more elegantly than I do. There cannot be a general specialist; the specialist can have no kind of authority, unless he has avowedly limited his range. There cannot be such a thing as the health adviser of the community, because there cannot be such a thing as one who specialises in the universe.

Thus when Dr. Saleeby says that a young man about to be married should be obliged to produce his health-book as he does his bank-book, the expression is neat; but it does not convey the real respects in which the two things agree, and in which they differ. To begin with, of course, there is a great deal too much of the bank-book for the sanity of our commonwealth; and it is highly probable that the health-book, as conducted in modern conditions, would rapidly become as timid, as snobbish, and as sterile as the money side of marriage has become. In the moral atmosphere of modernity the poor and the honest would probably get as much the worst of it if we fought with health-books as they do when we fight with bank-books. But that is a more general matter; the real point is in the difference between the two. The difference is in this vital fact: that a monied man generally thinks about money, whereas a healthy man does not think about health. If the strong young man cannot produce his health-book, it is for the perfectly simple reason that he has not got one. He can mention some extraordinary malady he has; but every man of honour is expected to do that now, whatever may be the decision that follows on the knowledge.

Health is simply Nature, and no naturalist ought to have the impudence to understand it. Health, one may say, is God; and no agnostic has any right to claim His acquaintance. For God must mean, among other things, that mystical and multitudinous balance of all things, by which they are at least able to stand up straight and endure; and any scientist who pretends to have exhausted this subject of ultimate sanity, I will call the lowest of religious fanatics. I will allow him to understand the madman, for the madman is an exception. But if he says he understands the sane man, then he says he has the secret of the Creator. For whenever you and I feel fully sane, we are quite incapable of naming the elements that make up that mysterious simplicity. We can no more analyse such peace in the soul than we can conceive in our heads the whole enormous and dizzy equilibrium by which, out of suns roaring like infernos and heavens toppling like precipices, He has hanged the world upon nothing.

We conclude, therefore, that unless Eugenic activity be restricted to monstrous things like mania, there is no constituted or constitutable authority that can really over-rule men in a matter in which they are so largely on a level. In the matter of fundamental human rights, nothing can be above Man, except God. An institution claiming to come from God might have such authority; but this is the last claim the Eugenists are likely to make. One caste or one profession seeking to rule men in such matters is like a man's right eye claiming to rule him, or his left leg to run away with him. It is madness. We now pass on to consider whether there is really anything in the way of Eugenics to be done, with such cheerfulness as we may possess after discovering that there is nobody to do it.

### Humanity Not Fallen

The sense of original sin would show, according to my theory, not that man was fallen from a high estate, but that he was rapidly rising from a low one. It would therefore confirm the conclusion that has been arrived at by every independent line of ethnological research—that our forefathers were utter savages from the beginning; and, that, after myriads of years of barbarism, our race has but recently grown to be civilized and religious.
—Francis Galton, 1865

### Higher and Lower Races

Another difference, which may either be due to natural selection or to original differences of race, is the fact that savages seem incapable of progress after the first few years of their life. The average children of all races are much on a par. Occasionally, those of the lower races are more precocious than the Anglo-Saxons; as a brute beast of a few weeks old is certainly more apt and forward than a child of the same age. But, as the years go by, the higher races continue to progress, while the lower ones gradually stop. They remain children in mind, with the passions of grown men.—Francis Galton, 1865

### Creating Long-Established Breeds

No one, I think, can doubt, from the facts and analogies I have brought forward, that, if talented men were mated with talented women, of the same mental and physical characters as themselves, generation after generation, we might produce a highly bred human race, with no more tendency to revert to meaner ancestral types than is shown by our long-established breeds of race-horses and foxhounds.—Francis Galton, 1865

# 6

# The Unanswered Challenge

One does not, as I have said, need to deny heredity in order to resist such legislation, any more than one needs to deny the spiritual world in order to resist an epidemic of witch-burning.

Dr. Saleeby did me the honour of referring to me in one of his addresses on this subject, and said that even I cannot produce any but a feeble-minded child from a feeble-minded ancestry. To which I reply, first of all, that he cannot produce a feeble-minded child. The whole point of our contention is that this phrase conveys nothing fixed and outside opinion. There is such a thing as mania, which has always been segregated; there is such a thing as idiotcy, which has always been segregated; but feeble-mindedness is a new phrase under which you might segregate anybody. It is essential that this fundamental fallacy in the use of statistics should be got somehow into the modern mind. Such people must be made to see the point, which is surely plain enough, that it is useless to have exact figures if they are exact figures about an inexact phrase. If I say, "There are five fools in Acton," it is surely quite clear that, though no mathematician can make five the same as four or six, that will not stop you or anyone else from finding a few more fools in Acton. Now weak-mindedness, like folly, is a term divided from madness in this vital manner—that in one sense it applies to all men, in another to most men, in another to very many men, and so on. It is as if Dr. Saleeby were to say, "Vanity, I find, is undoubtedly hereditary. Here is Mrs. Jones, who was very sensitive about her sonnets being criticised, and I found her little daughter in a new frock looking in the glass. The experiment is conclusive, the demonstration is complete; there in the first generation is the artistic temperament—that is vanity; and there in the second generation is dress—and that is vanity."

We should answer, "My friend, all is vanity, vanity and vexation of spirit—especially when one has to listen to logic of your favourite kind. Obviously all human beings must value themselves; and obviously there is in all such valuation an element of weakness, since it is not the valuation of eternal justice. What is the use of your finding by experiment in some people a thing we know by reason must be in all of them?"

Here it will be as well to pause a moment and avert one possible misunderstanding. I do not mean that you and I cannot and do not practically see and personally remark on this or that eccentric or intermediate type, for which the word "feeble-minded" might be a very convenient word, and might correspond to a genuine though indefinable fact of experience. In the same way we might speak, and do speak, of such and such a person being "mad with vanity" without wanting two keepers to walk in and take the person off. But I ask the reader to remember always that I am talking of words, not as they are used in talk or novels, but as they will be used, and have been used, in warrants and certificates, and Acts of Parliament. The distinction between the two is perfectly clear and practical. The difference is that a novelist or a talker can be trusted to try and hit the mark; it is all to his glory that the cap should fit, that the type should be recognised; that he should, in a literary sense, hang the right man. But it is by no means always to the interests of governments or officials to hang the right man. The fact that they often do stretch words in order to cover cases is the whole foundation of having any fixed laws or free institutions at all. My point is not that I have never met anyone whom I should call feeble-minded, rather than mad or imbecile. My point is that if I want to dispossess a nephew, oust a rival, silence a blackmailer, or get rid of an importunate widow, there is nothing in logic to prevent my calling them feeble-minded too. And the vaguer the charge is the less they will be able to disprove it.

One does not, as I have said, need to deny heredity in order to resist such legislation, any more than one needs to deny the spiritual world in order to resist an epidemic of witch-burning. I admit there may be such a thing as hereditary feeble-mindedness; I believe there is such a thing as witchcraft. Believing that there are spirits, I am bound in mere reason to suppose that there are probably evil spirits; believing that there are evil spirits, I am bound in mere reason to suppose that some men grow evil by dealing with them. All that is mere rationalism; the superstition (that is the unreasoning repugnance and terror) is in the person who admits there can be angels but denies there can be devils. The superstition is in the person who admits there can be devils but denies there can be diabolists. Yet I should certainly resist any effort to search for witches, for a perfectly simple reason, which is the key of the whole of this controversy. The reason is that it is one thing to believe in witches, and quite another to believe in witch-smellers. I have more respect for the old witch-finders than for the Eugenists, who go about persecuting the fool of the family; because the witch-finders according to their own conviction, ran a risk. Witches were not the feeble-minded, but the strong-minded—the evil mesmerists, the rulers of the elements. Many a raid on a witch, right or wrong, seemed to the villagers who did it a righteous popular rising against a vast spiritual tyranny, a papacy of sin. Yet we know that the thing degenerated into a rabid and despicable persecution of the feeble or the old. It ended by being a war upon the weak. It ended by being what Eugenics begins by being.

### The 'Better Dead' School of Eugenics

That is why I define negative eugenics as "the discouragement of unworthy *Parenthood*," a project which involves no killing, and is morally at the opposite pole from that of the purely "Darwinian" eugenists, who advocate a return to natural selection with its destruction of the unfortunate, and whom I define as the "better-dead" school of eugenists. For that is what so-called "Darwinism" leads to: the championship of infant mortality, contempt for mercy, enmity to social reform, and the prostitution of divine eugenics to the diabolical creed of "Each man for himself, and the devil take the hindmost."—C.W. SALEEBY, 1914

---

When I said above that I believed in witches, but not in witch-smellers, I stated my full position about that conception of heredity, that half-formed philosophy of fears and omens; of curses and weird recurrence and darkness and the doom of blood, which, as preached to humanity to-day, is often more inhuman than witchcraft itself. I do not deny that this dark element exists; I only affirm that it is dark; or, in other words, that its most strenuous students are evidently in the dark about it. I would no more trust Dr. Karl Pearson on a heredity-hunt than on a heresy-hunt. I am perfectly ready to give my reasons for thinking this; and I believe any well-balanced person, if he reflects on them, will think as I do. There are two senses in which a man may be said to know or not know a subject. I know the subject of arithmetic, for instance; that is, I am not good at it, but I know what it is. I am sufficiently familiar with its use to see the absurdity of anyone who says, "So vulgar a fraction cannot be mentioned before ladies," or "This unit is Unionist, I hope." Considering myself for one moment as an arithmetician, I may say that I know next to nothing about my subject: but I know my subject. I know it in the street. There is the other kind of man, like Dr. Karl Pearson, who undoubtedly knows a vast amount about his subject; who undoubtedly lives in great forests of facts concerning kinship and inheritance. But it is not, by any means, the same thing to have searched the forests and to have recognised the frontiers. Indeed, the two things generally belong to two very different types of mind. I gravely doubt whether the Astronomer-Royal would write the best essay on the relations between astronomy and astrology. I doubt whether the President of the Geographical Society could give the best definition and history of the words "geography" and "geology."

Now the students of heredity, especially, understand all of their subject except their subject. They were, I suppose, bred and born in that brier-patch, and have really explored it without coming to the end of it. That is, they have studied everything but the question of what they are studying. Now I do not propose to rely merely on myself to tell them what they are studying. I propose, as will be seen in a moment, to call the testimony of a great man who has himself studied it. But to begin with, the domain of heredity (for those who see its frontiers) is a sort of triangle, enclosed on its three sides by three facts. The first is that heredity undoubtedly exists, or there would be no such thing as a family likeness, and every marriage might suddenly produce a small negro. The second is that even simple heredity can never be simple; its complexity must be literally unfathomable, for in that field fight unthinkable millions. But yet again it never is simple heredity: for the instant anyone is, he experiences. The third is that these innumerable ancient influences, these instant inundations of experiences, come together according to a combination that is unlike anything else on this earth. It is a combination that does combine. It cannot be sorted out again, even on the Day of Judgment. Two totally different people have become in the sense most sacred, frightful, and unanswerable, one flesh. If a golden-haired Scandinavian girl has married a very swarthy Jew, the Scandinavian side of the family may say till they are blue in the face that the baby has his mother's nose or his mother's eyes. They can never be certain the black-haired Bedouin is not present in every feature, in every inch. In the person of the baby he may have gently pulled his wife's nose. In the person of the baby he may have partly blacked his wife's eyes.

Those are the three first facts of heredity. That it exists; that it is subtle and made of a million elements; that it is simple, and cannot be unmade into those elements. To summarise: you know there is wine in the soup. You do not know how many wines there are in the soup, because you do not know how many wines there are in the world. And you never will know, because all chemists, all cooks, and all common-sense people tell you that the soup is of such a sort that it can never be chemically analysed. That is a perfectly fair parallel to the hereditary element in the human soul. There are many ways in which one can feel that there is wine in the soup, as in suddenly tasting a wine specially favoured; that corresponds to seeing suddenly flash on a young face the image of some ancestor you have known. But even then the taster cannot be certain he is not tasting one familiar wine among many unfamiliar ones—or seeing one known ancestor among a million unknown ancestors. Another way is to get drunk on the soup, which corresponds to the case of those who say they are driven to sin and death by hereditary doom. But even then the drunkard cannot be certain it was the soup, any more than the traditional drunkard who is certain it was the salmon.

Those are the facts about heredity which anyone can see. The upshot of them is not only that a miss is as good as a mile, but a miss is as good as a win. If the child has his parents' nose (or noses) that may be heredity. But if he has not, that may be heredity too. And as we need not take heredity lightly because two generations differ—so we need not take heredity a scrap more seriously because two generations are similar. The thing is there, in what cases we know not, in what proportion we know not, and we cannot know.

Now it is just here that the decent difference of function between Dr. Saleeby's trade and mine comes in. It is his business to study human health and sickness as a whole, in a spirit of more or less enlightened guesswork; and it is perfectly natural that he should allow for heredity here, there, and everywhere as a man climbing a mountain or sailing a boat will allow for weather without even explaining it to himself. An utterly different attitude is incumbent on any conscientious man writing about what laws should be enforced or about how commonwealths should be governed. And when we consider how plain a fact is murder, and yet how hesitant and even hazy we all grow about the guilt of a murderer, when we consider how simple an act is stealing, and yet how hard it is to convict and punish those rich commercial pirates who steal the most, when we consider how cruel and clumsy the law can be even about things as old and plain as the Ten Commandments—I simply cannot conceive any responsible person proposing to legislate on our broken knowledge and bottomless ignorance of heredity.

But though I have to consider this dull matter in its due logical order, it appears to me that this part of the matter has been settled, and settled in a most masterly way, by somebody who has infinitely more right to speak on it than I have. Our press seems to have a perfect genius for fitting people with caps that don't fit; and affixing the wrong terms of eulogy and even the wrong terms of abuse. And just as people will talk of Bernard Shaw as a naughty winking Pierrot, when he is the last great Puritan and really believes in respectability; just as (*si parva licet* etc.) they will talk of my own paradoxes, when I pass my life in preaching that the truisms are true; so an enormous number of newspaper readers seem to have it fixed firmly in their heads that Mr. H. G. Wells is a harsh and horrible Eugenist in great goblin spectacles, who wants to put us all into metallic microscopes and dissect us with metallic tools. As a matter of fact, of course, Mr. Wells, so far from being too definite, is generally not definite enough. He is an absolute wizard in the appreciation of atmospheres and the opening of vistas; but his answers are more agnostic than his questions. His books will do every thing except shut. And so far from being the sort of man who would stop a man from propagating, he cannot even stop a full stop. He is not Eugenic enough to prevent the black dot at the end of a sentence from breeding a line of little dots.

But this is not the clear-cut blunder of which I spoke. The real blunder is this. Mr. Wells deserves a tiara of crowns and a garland of medals for all kinds of reasons. But if I were restricted, on grounds of public economy, to giving Mr. Wells only one medal *ob cives servatos,* I would give him a medal as the Eugenist who destroyed Eugenics. For everyone spoke of him, rightly or wrongly, as a Eugenist; and he certainly had, as I have not, the training and type of culture required to consider the matter merely in a biological and not in a generally moral sense. The result was that in that fine book, *Mankind in the Making,* where he inevitably came to grips with the problem, he threw down to the Eugenists an intellectual challenge which seems to me unanswerable, but which, at any rate, is unanswered. I do not mean that no remote Eugenist wrote upon the subject; for it is impossible to read all writings, especially Eugenist writings. I do mean that the leading Eugenists write as if this challenge had never been offered. The gauntlet lies unlifted on the ground.

Having given honour for the idea where it is due, I may be permitted to summarise it myself for the sake of brevity. Mr. Wells' point was this. That we cannot be certain about the inheritance of health, because health is not a quality. It is not a thing like darkness in the hair or length in the limbs. It is a relation, a balance. You have a tall, strong man; but his very strength depends on his not being too tall for his strength. You catch a healthy, full-blooded fellow; but his very health depends on his being not too full of blood. A heart that is strong for a dwarf will be weak for a giant; a nervous system that would kill a man with a trace of a certain illness will sustain him to ninety if he has no trace of that illness. Nay, the same nervous system might kill him if he had an excess of some other comparatively healthy thing. Seeing, therefore, that there are apparently healthy people of all types, it is obvious that if you mate two of them, you may even then produce a discord out of two inconsistent harmonies. It is obvious that you can no more be certain of a good offspring than you can be certain of a good tune if you play two fine airs at once on the same piano. You can be even less certain of it in the more delicate case of beauty, of which the Eugenists talk a great deal. Marry two handsome people whose noses tend to the aquiline, and their baby (for all you know) may be a goblin with a nose like an enormous parrot's. Indeed, I actually know a case of this kind. The Eugenist has to settle, not the result of fixing one steady thing to a second steady thing; but what will happen when one toppling and dizzy equilibrium crashes into another.

This is the interesting conclusion. It is on this degree of knowledge that we are asked to abandon the universal morality of mankind. When we have stopped the lover from marrying the unfortunate woman he loves, when we have found him another uproariously healthy female whom he does not love in the least, even then we have no logical evidence that the result may not be as horrid and dangerous as if he had behaved like a man of honour.

### Targeting Asians and Eastern Europeans with Birth Control

This brings home to one the importance of your work in C.B.C. [Constructive Birth Control] and of lessening the number of births from inferior human parents, and those unable to maintain their children, in our islands.

But one feels that unless at the same time the influx of low-caste foreigners, especially from Eastern Europe, is checked, they will fill up the gaps and mongrelized our English and Scotch stock.

Like the rats, these low-caste foreigners have large families, and are industrious workers and have strong tribal instincts, but compared with our people they are cunning, bloodthirsty and cowardly.

In Eastern Europe and Asia large families are the insurance for old age. The more children they have the better will the parents be supported, it being considered the old and elderly parents right to live on their children. So, instead of saving, they have as many children as possible. Their priests encourage them as it increases the number of their "faithful."

Very likely the Society has this race question in hand. A few strains of strong races like the French (as the Huguenots) or Scandinavians do good in a pedigree and blend well with us, but these Eastern and Eastern Europe races are altogether inferior to ours.

The London working-classes are distinctly more foreign-looking than they were in my childhood, and we are losing our advantage of living on an island.—LETTER TO *BIRTH CONTROL NEWS*, NOV, 1922

CHAPTER

7

# The Established Church
# of Doubt

The thing that really is trying to tyrannise through government is Science. The thing that really does use the secular arm is Science. And the
creed that really is levying tithes and capturing schools, the creed that
really is enforced by fine and imprisonment, the creed that really is proclaimed not in sermons but in statutes, and spread not by pilgrims but by
policemen—that creed is the great but disputed system of thought which
began with Evolution and has ended in Eugenics.

Let us now finally consider what the honest Eugenists do mean, since it
has become increasingly evident that they cannot mean what they say.
Unfortunately, the obstacles to any explanation of this are such as to insist
on a circuitous approach. The tendency of all that is printed and much that
is spoken to-day is to be, in the only true sense, behind the times. It is
because it is always in a hurry that it is always too late. Give an ordinary
man a day to write an article, and he will remember the things he has
really heard latest; and may even, in the last glory of the sunset, begin to
think of what he thinks himself. Give him an hour to write it, and he will
think of the nearest text-book on the topic, and make the best mosaic he
may out of classical quotations and old authorities. Give him ten minutes
to write it and he will run screaming for refuge to the old nursery where
he learnt his stalest proverbs, or the old school where he learnt his stalest
politics. The quicker goes the journalist the slower go his thoughts. The
result is the newspaper of our time, which every day can be delivered earlier and earlier, and which, every day is less worth delivering at all. The
poor panting critic falls farther and farther behind the motor-car of modern fact. Fifty years ago he was barely fifteen years behind the times. Fifteen years ago he was not more than fifty years behind the times. Just now
he is rather more than a hundred years behind the times: and the proof of

it is that the things he says, though manifest nonsense about our society to-day, really were true about our society some hundred and thirty years ago. The best instance of his belated state is his perpetual assertion that the supernatural is less and less believed. It is a perfectly true and realistic account—of the eighteenth century. It is the worst possible account of this age of psychics and spirit-healers and fakirs and fashionable fortune-tellers. In fact, I generally reply in eighteenth century language to this eighteenth century illusion. If somebody says to me, "The creeds are crumbling," I reply, "And the King of Prussia, who is himself a Freethinker, is certainly capturing Silesia from the Catholic Empress." If somebody says, "Miracles must be reconsidered in the light of rational experience," I answer affably, "But I hope that our enlightened leader, Hébert will not insist on guillotining that poor French queen." If somebody says, "We must watch for the rise of some new religion which can commend itself to reason," I reply, "But how much more necessary is it to watch for the rise of some military adventurer who may destroy the Republic; and, to my mind, that young Major Bonaparte has rather a restless air." It is only in such language from the Age of Reason that we can answer such things. The age we live in is something more than an age of superstition—it is an age of innumerable superstitions. But it is only with one example of this that I am concerned here.

### Eugenics as Religion

[Eugenics] must be introduced into the national conscience, like a new religion. It has, indeed, strong claims to become an orthodox religious tenet of the future, for eugenics co-operate with the workings of nature by securing that humanity shall be represented by the fittest races.—FRANCIS GALTON, 1904

I mean the error that still sends men marching about disestablishing churches and talking of the tyranny of compulsory church teaching or compulsory church tithes. I do not wish for an irrelevant misunderstanding here; I would myself certainly disestablish any church that had a numerical minority, like the Irish or the Welsh, and I think it would do a great deal of good to genuine churches that have a partly conventional majority, like the English, or even the Russian. But I should only do this if I had nothing else to do; and just now there is very much else to do. For religion, orthodox or unorthodox, is not just now relying on the weapon of State establishment at all. The Pope practically made no attempt to preserve the Concordat; but seemed rather relieved at the independence his Church gained by the destruction of it: and it is common talk among the French clericalists that the Church has gained by the change. In Russia the one real charge brought by religious people (especially Roman Catholics) against the Orthodox Church is not its orthodoxy or heterodoxy, but its abject dependence on the State. In England we can almost measure an Anglican's fervour for his Church by his comparative coolness about its establishment—that is, its control by a Parliament of Scotch Presbyterians like Balfour, or Welsh Congregationalists like Lloyd George. In Scotland the powerful combination of the two great sects outside the establishment have left it in a position in which it feels no disposition to boast of being called by mere lawyers the Church of Scotland. I am not here arguing that Churches should not depend on the State; nor that they do not depend upon much worse things. It may be reasonably maintained that the strength of Romanism, though it be not in any national police, is in a moral police more rigid and vigilant. It may be reasonably maintained that the strength of Anglicanism, though it be not in establishment, is in aristocracy, and its shadow, which is called snobbishness. All I assert here is that the Churches are not now leaning heavily on their political establishment; they are not using heavily the secular arm. Almost everywhere their legal tithes have been modified, their legal boards of control have been mixed. They may still employ tyranny, and worse tyranny: I am not considering that. They are not specially using that special tyranny which consists in using the government.

The thing that really is trying to tyrannise through government is Science. The thing that really does use the secular arm is Science. And the creed that really is levying tithes and capturing schools, the creed that really is enforced by fine and imprisonment, the creed that really is proclaimed not in sermons but in statutes, and spread not by pilgrims but by policemen—that creed is the great but disputed system of thought which began with Evolution and has ended in Eugenics. Materialism is really our established Church; for the Government will really help it to persecute its heretics. Vaccination, in its hundred years of experiment, has been disputed almost as much as baptism in its approximate two thousand. But it seems quite natural to our politicians to enforce vaccination; and it would seem to them madness to enforce baptism.

### Science Leads a Revolution

Professor Vallon mentions, without sympathy, the hysterical denunciations of Eugenics by Messrs. Belloc and Chesterton. We may wonder why these popular writers and journalists should wish to fill England with degenerates. But they have a reason for their incoherent rage. The realise that Science, instead of confining itself to making bad smells in laboratories, is calmly preparing to lead a social and moral revolution, a revolution in which neither medieval casuistry nor Marxian class-war will count for anything at all. The great struggle of the future will be between Science and its enemies.

—WILLIAM INGE, *EUGENICS REVIEW*, 1924

I am not frightened of the word "persecution" when it is attributed to the churches; nor is it in the least as a term of reproach that I attribute it to the men of science. It is as a term of legal fact. If it means the imposition by the police of a widely disputed theory, incapable of final proof—then our priests are not now persecuting, but our doctors are. The imposition of such dogmas constitutes a State Church—in an older and stronger sense than any that can be applied to any supernatural Church to-day. There are still places where the religious minority is forbidden to assemble or to teach in this way or that; and yet more where it is excluded from this or that Public Post. But I cannot now recall any place where it is compelled by the criminal law to go through the rite of the official religion. Even the Young Turks did not insist on all Macedonians being circumcised.

### Eugenics Will Save Civilization

I agree with the paper, and go so far as to say that there is now no reasonable excuse for refusing to face the fact that nothing but a eugenic religion can save our civilisation from the fate that has overtaken all previous civilisations.
—George Bernard Shaw, 1904

Now here we find ourselves confronted with an amazing fact. When, in the past, opinions so arguable have been enforced by State violence, it has been at the instigation of fanatics who held them for fixed and flaming certainties. If truths could not be evaded by their enemies, neither could they be altered even by their friends. But what are the certain truths that the secular arm must now lift the sword to enforce? Why, they are that very mass of bottomless questions and bewildered answers that we have been studying in the last chapters—questions whose only interest is that they are trackless and mysterious; answers whose only glory is that they are tentative and new. The devotee boasted that he would never abandon the faith; and therefore he persecuted for the faith. But the doctor of science actually boasts that he will always abandon a hypothesis; and yet he persecutes for the hypothesis. The Inquisitor violently enforced his creed, because it was unchangeable. The *savant* enforces it violently because he may change it the next day.

Now this is a new sort of persecution; and one may be permitted to ask if it is an improvement on the old. The difference, so far as one can see at first, seems rather favourable to the old. If we are to be at the merciless mercy of man, most of us would rather be racked for a creed that existed intensely in somebody's head, rather than vivisected for a discovery that had not yet come into anyone's head, and possibly never would. A man would rather be tortured with a thumbscrew until he chose to see reason than tortured with a vivisecting knife until the vivisector chose to see reason. Yet that is the real difference between the two types of legal enforcement. If I gave in to the Inquisitors, I should at least know what creed to profess. But even if I yelled out *a credo* when the Eugenists had me on the rack, I should not know what creed to yell. I might get an extra turn of the rack for confessing to the creed they confessed quite a week ago.

Now let no light-minded person say that I am here taking extravagant parallels; for the parallel is not only perfect, but plain. For this reason: that the difference between torture and vivisection is not in any way affected by the fierceness or mildness of either. Whether they gave the rack half a turn or half a hundred, they were, by hypothesis, dealing with a truth which they knew to be there. Whether they vivisect painfully or painlessly, they are trying to find out whether the truth is there or not. The old Inquisitors tortured to put their own opinions into somebody. But the new Inquisitors torture to get their own opinions out of him. They do not know what their own opinions are, until the victim of vivisection tells them. The division of thought is a complete chasm for anyone who cares about thinking. The old persecutor was trying to *teach* the citizen, with fire and sword. The new persecutor is trying to learn from the citizen, with scalpel and germ-injector. The master was meeker than the pupil will be.

### Ruthless Sterilization

In my opinion there is only one remedy for this state of affairs, and that is the ruthless sterilization of the mental defective so that they may not be able to hand on these defects to posterity.
—PROF. E.W. MACBRIDE, *BIRTH CONTROL NEWS*, FEB. 1923

I could prove by many practical instances that even my illustrations are not exaggerated, by many placid proposals I have heard for the vivisection of criminals, or by the filthy incident of Dr. Neisser. But I prefer here to stick to a strictly logical line of distinction, and insist that whereas in all previous persecutions the violence was used to end our indecision, the whole point here is that the violence is used to end the indecision of the persecutors. This is what the honest Eugenists really mean, so far as they mean anything. They mean that the public is to be given up, not as a heathen land for conversion, but simply as a *pabulum* for experiment. That is the real, rude, barbaric sense behind this Eugenic legislation. The Eugenist doctors are not such fools as they look in the light of any logical inquiry about what they want. They do not know what they want, except that they want your soul and body and mine in order to find out. They are quite seriously, as they themselves might say, the first religion to be experimental instead of doctrinal. All other established Churches have been based on somebody having found the truth. This is the first Church that was ever based on not having found it.

### Eugenics as Religious Dogma

We are ignorant of the ultimate destinies of humanity, but feel perfectly sure that it is as noble a work to raise its level, in the sense already explained, as it would be disgraceful to abase it. I see no impossibility in eugenics becoming a religious dogma among mankind, but its details must first be worked out sedulously in the study.—FRANCIS GALTON, 1904

There is in them a perfectly sincere hope and enthusiasm; but it is not for us, but for what they might learn from us, if they could rule us as they can rabbits. They cannot tell us anything about heredity, because they do not know anything about it. But they do quite honestly believe that they would know something about it, when they had married and mismarried us for a few hundred years. They cannot tell us who is fit to wield such authority, for they know that nobody is; but they do quite honestly believe that when that authority has been abused for a very long time, somebody somehow will be evolved who is fit for the job. I am no Puritan, and no one who knows my opinions will consider it a mere criminal charge if I say that they are simply gambling. The reckless gambler has no money in his pockets; he has only the ideas in his head. These gamblers have no ideas in their heads; they have only the money in their pockets. But they think that if they could use the money to buy a big society to experiment on, something like an idea might come to them at last. That is Eugenics.

I confine myself here to remarking that I do not like it. I may be very stingy, but I am willing to pay the scientist for what he does know; I draw the line at paying him for everything he doesn't know. I may be very cowardly, but I am willing to be hurt for what I think or what he thinks—I am not willing to be hurt, or even inconvenienced, for whatever he might happen to think after he had hurt me. The ordinary citizen may easily be more magnanimous than I, and take the whole thing on trust; in which case his career may be happier in the next world, but (I think) sadder in this. At least, I wish to point out to him that he will not be giving his glorious body as soldiers give it, to the glory of a fixed flag, or martyrs to the glory of a deathless God. He will be, in the strict sense of the Latin phrase, giving his vile body for an experiment—an experiment of which even the experimentalist knows neither the significance nor the end.

# 8

# A Summary of a False Theory

There is no reason in Eugenics, but there is plenty of motive. Its supporters are highly vague about its theory, but they will be painfully practical about its practice. And while I reiterate that many of its more eloquent agents are probably quite innocent instruments, there *are* some, even among Eugenists, who by this time know what they are doing.

I have up to this point treated the Eugenists, I hope, as seriously as they treat themselves. I have attempted an analysis of their theory as if it were an utterly abstract and disinterested theory; and so considered, there seems to be very little left of it. But before I go on, in the second part of this book, to talk of the ugly things that really are left, I wish to recapitulate the essential points in their essential order, lest any personal irrelevance or over-emphasis (to which I know myself to be prone) should have confused the course of what I believe to be a perfectly fair and consistent argument. To make it yet clearer, I will summarise the thing under chapters, and in quite short paragraphs.

In the first chapter I attempted to define the essential point in which Eugenics can claim, and does claim, to be a new morality. That point is that it is possible to consider the baby in considering the bride. I do not adopt the ideal irresponsibility of the man who said, "What has posterity done for us?" But I do say, to start with, "What can we do for posterity, except deal fairly with our contemporaries?" Unless a man love his wife whom he has seen, how shall he love his child whom he has not seen?

In the second chapter I point out that this division in the conscience cannot be met by mere mental confusions, which would make any woman refusing any man a Eugenist. There will always be something in the world which tends to keep outrageous unions exceptional; that influence is not Eugenics, but laughter.

In the third chapter I seek to describe the quite extraordinary atmosphere in which such things have become possible. I call that atmosphere anarchy; but insist that it is an anarchy in the centres where there should be authority. Government has become ungovernable; that is, it cannot leave off governing. Law has become lawless; that is, it cannot see where laws should stop. The chief feature of our time is the meekness of the mob and the madness of the government. In this atmosphere it is natural enough that medical experts, being authorities, should go mad, and attempt so crude and random and immature a dream as this of petting and patting (and rather spoiling) the babe unborn.

In chapter four I point out how this impatience has burst through the narrow channel of the Lunacy Laws, and has obliterated them by extending them. The whole point of the madman is that he is the exception that proves the rule. But Eugenics seeks to treat the whole rule as a series of exceptions—to make all men mad. And on that ground there is hope for nobody; for all opinions have an author, and all authors have a heredity. The mentality of the Eugenist makes him believe in Eugenics as much as the mentality of the reckless lover makes him violate Eugenics and both mentalities are, on the materialist hypothesis, equally the irresponsible product of more or less unknown physical causes. The real security of man against any logical Eugenics is like the false security of Macbeth. The only Eugenist that could rationally attack him must be a man of no woman born.

In the chapter following this, which is called "The Flying Authority," I try in vain to locate and fix any authority that could rationally rule men in so rooted and universal a matter; little would be gained by ordinary men doing it to each other; and if ordinary practitioners did it they would very soon show, by a thousand whims and quarrels, that they were ordinary men. I then discussed the enlightened despotism of a few general professors of hygiene, and found it unworkable, for an essential reason: that while we can always get men intelligent enough to know more than the rest of us about this or that accident or pain or pest, we cannot count on the appearance of great cosmic philosophers; and only such men can be even supposed to know more than we do about normal conduct and common sanity. Every sort of man, in short, would shirk such a responsibility, except the worst sort of man, who would accept it.

I pass on, in the next chapter, to consider whether we know enough about heredity to act decisively, even if we were certain who ought to act.

Here I refer the Eugenists to the reply of Mr. Wells, which they have never dealt with to my knowledge or satisfaction—the important and primary objection that health is not a quality but a proportion of qualities; so that even health married to health might produce the exaggeration called disease. It should be noted here, of course, that an individual biologist may quite honestly believe that he has found a fixed principle with the help of Weissmann or Mendel. But we are not discussing whether he knows enough to be justified in thinking (as is somewhat the habit of the anthropoid *Homo*) that he is right. We are discussing whether *we* know enough, as responsible citizens, to put such powers into the hands of men who may be deceived or who may be deceivers. I conclude that we do not.

In the last chapter of the first half of the book I give what is, I believe, the real secret of this confusion, the secret of what the Eugenists really want. They want to be allowed to find out what they want. Not content with the endowment of research, they desire the establishment of research; that is the making of it a thing official and compulsory, like education or state insurance; but still it is only research and not discovery. In short, they want a new kind of State Church, which shall be an Established Church of Doubt—instead of Faith. They have no Science of Eugenics at all, but they do really mean that if we will give ourselves up to be vivisected they may very probably have one some day. I point out, in more dignified diction, that this is a bit thick.

And now, in the second half of this book, we will proceed to the consideration of things that really exist. It is, I deeply regret to say, necessary to return to realities, as they are in your daily life and mine. Our happy holiday in the land of nonsense is over; we shall see no more its beautiful city, with the almost Biblical name of Bosh, nor the forests full of mares' nests, nor the fields of tares that are ripened only by moonshine. We shall meet no longer those delicious monsters that might have talked in the same wild club with the Snark and the Jabberwock or the Pobble or the Dong with the Luminous Nose; the father who can't make head or tail of the mother, but thoroughly understands the child she will some day bear; the lawyer who has to run after his own laws almost as fast as the criminals run away from them; the two mad doctors who might discuss for a million years which of them has the right to lock up the other; the grammarian who clings convulsively to the Passive Mood, and says it is the duty of something to get itself done without any human assistance; the man who would marry giants to giants until the back breaks, as children pile brick upon brick for the pleasure of seeing the staggering tower tumble down; and, above all, the superb man of science who wants you to pay him and crown him because he has so far found out nothing. These fairy-tale comrades must leave us. They exist, but they have no influence in what is really going on. They are honest dupes and tools, as you and I were very nearly being honest dupes and tools. If we come to think coolly of the world we live in, if we consider how very practical is the practical politician, at least where cash is concerned, how very dull and earthy are most of the men who own the millions and manage the newspaper trusts, how very cautious and averse from idealist upheaval are those that control this capitalist society—when we consider all this, it is frankly incredible that Eugenics should be a front bench fashionable topic and almost an Act of Parliament, if it were in practice only the unfinished fantasy which it is, as I have shown, in pure reason. Even if it were a just revolution, it would be much too revolutionary a revolution for modern statesmen, if there were not something else behind. Even if it were a true ideal, it would be much too idealistic an ideal for our "practical men," if there were not something real as well. Well, there is something real as well. There is no reason in Eugenics, but there is plenty of motive. Its supporters are highly vague about its theory, but they will be painfully practical about its practice. And while I reiterate that many of its more eloquent agents are probably quite innocent instruments, there *are* some, even among Eugenists, who by this time know what they are doing. To them we shall not say, "What is Eugenics?" or "Where on earth are you going?" but only "Woe Unto you, hypocrites, that devour widows' houses and for a pretence use long words."

### Natural Selection versus Eugenics

Natural selection and eugenic selection may have the same effect and end, but they are fundamentally distinct in method. Natural selection is a selective death-rate, killing those less able to survive, but eugenic selection, in Professor Pearson's own admirable phrase, replaces this selective death-rate by a selective birth-rate: and no form of killing or permission of killing can be anything but a negation of the essential characteristic of eugenics.

—C.W. SALEEBY, 1914

# 9

# The Impotence of Impenitence

At the beginning of our epoch men talked with equal ease about Reform and Repeal. Now everybody talks about reform; but nobody talks about repeal. Our fathers did not talk of Free Trade, but of the Repeal of the Corn Laws. They did not talk of Home Rule, but of the Repeal of the Union. In those days people talked of a "Repealer" as the most practical of all politicians, the kind of politician that carries a club. Now the Repealer is flung far into the province of an impossible idealism: and the leader of one of our great parties, having said, in a heat of temporary sincerity, that he would repeal an Act, actually had to write to all the papers to assure them that he would only amend it.

The root formula of an epoch is always an unwritten law, just as the law that is the first of all laws, that which protects life from the murderer, is written nowhere in the Statute Book. Nevertheless there is all the difference between having and not having a notion of this basic assumption in an epoch. For instance, the Middle Ages will simply puzzle us with their charities and cruelties, their asceticism and bright colours, unless we catch their general eagerness, for building and planning, dividing this from that by walls and fences—the spirit that made architecture their most successful art. Thus even a slave seemed sacred; the divinity that did hedge a king, did also, in one sense, hedge a serf, for he could not be driven out from behind his hedges. Thus even liberty became a positive thing like a privilege; and even, when most men had it, it was not opened like the freedom of a wilderness, but bestowed, like the freedom of a city. Or again, the seventeenth century may seem a chaos of contradictions, with its almost priggish praise of parliaments and its quite barbaric massacre of prisoners, until we realise that, if the Middle Ages was a house half built, the seventeenth century was a house on fire. Panic was the note of it, and that fierce fastidiousness and exclusiveness that comes from fear. Calvinism was its characteristic religion, even in the Catholic Church, the insistence on the narrowness of the way and the fewness of the chosen. Suspicion was the note of its politics—"put not your trust in princes." It tried to thrash everything out by learned, virulent, and ceaseless controversy; and it weeded its population by witch-burning. Or yet again: the eighteenth century will present pictures that seem utterly opposite, and yet seem singularly typical of the time: the sack of Versailles and the *Vicar of Wakefield;* the pastorals of Watteau and the dynamite speeches of Danton. But we shall understand them all better if we once catch sight of the idea of *tidying up* which ran through the whole period, the quietest people being prouder of their tidiness, civilisation, and sound taste than of any of their virtues; and the wildest people having (and this is the most important point) no love of wildness for its own sake, like Nietzsche or the anarchic poets, but only a readiness to employ it to get rid of unreason or disorder. With these epochs it is not altogether impossible to say that some such form of words is a key. The epoch for which it is almost impossible to find a form of words is our own.

Nevertheless, I think that with us the keyword is "inevitability," or, as I should be inclined to call it, "impenitence." We are subconsciously dominated in all departments by the notion that there is no turning back, and it is rooted in materialism and the denial of free-will. Take any handful of modern facts and compare them with the corresponding facts a few hundred years ago. Compare the modern Party System with the political factions of the seventeenth century. The difference is that in the older time the party leaders not only really cut off each other's heads, but (what is much more alarming) really repealed each other's laws. With us it has become traditional for one party to inherit and leave untouched the acts of the other when made, however bitterly they were attacked in the making. James II and his nephew William were neither of them very gay specimens; but they would both have laughed at the idea of "a continuous foreign policy." The Tories were not Conservatives; they were, in the literal sense, reactionaries. They did not merely want to keep the Stuarts; they wanted to bring them back.

Or again, consider how obstinately the English mediaeval monarchy returned again and again to its vision of French possessions, trying to reverse the decision of fate; how Edward III returned to the charge after the defeats of John and Henry III, and Henry V after the failure of Edward III; and how even Mary had that written on her heart which was neither her husband nor her religion. And then consider this: that we have comparatively lately known a universal orgy of the thing called Imperialism, the unity of the Empire the only topic, colonies counted like crown jewels, and the Union Jack waved across the world. And yet no one so much as dreamed, I will not say of recovering, the American colonies for the Imperial unity (which would have been too dangerous a task for modern empire-builders), but even of re-telling the story from an Imperial standpoint. Henry V justified the claims of Edward III. Joseph Chamberlain would not have dreamed of justifying the claims of George III. Nay, Shakespeare justifies the French War, and sticks to Talbot and defies the legend of Joan of Arc. Mr. Kipling would not dare to justify the American War, stick to Burgoyne, and defy the legend of Washington. Yet there really was much more to be said for George III than there ever was for Henry V. It was not said, much less acted upon, by the modern Imperialists; because of this basic modern sense, that as the future is inevitable, so is the past irrevocable. Any fact so complete as the American exodus from the Empire must be considered as final for aeons, though it hardly happened more than a hundred years ago. Merely because it has managed to occur it must be called first, a necessary evil, and then an indispensable good. I need not add that I do not want to reconquer America; but then I am not an Imperialist.

Then there is another way of testing it: ask yourself how many people you have met who grumbled at a thing as incurable, and how many who attacked it as curable? How many people we have heard abuse the British elementary schools, as they would abuse the British climate? How few have we met who realised that British education can be altered, but British weather cannot? How few there were that knew that the clouds were more immortal and more solid than the schools? For a thousand that regret compulsory education, where is the hundred, or the ten, or the one, who would repeal compulsory education? Indeed, the very word proves my case by its unpromising and unfamiliar sound. At the beginning of our epoch men talked with equal ease about Reform and Repeal. Now everybody talks about reform; but nobody talks about repeal. Our fathers did not talk of Free Trade, but of the Repeal of the Corn Laws. They did not talk of Home Rule, but of the Repeal of the Union. In those days people talked of a "Repealer" as the most practical of all politicians, the kind of politician that carries a club. Now the Repealer is flung far into the province of an impossible idealism: and the leader of one of our great parties, having said, in a heat of temporary sincerity, that he would repeal an Act, actually had to write to all the papers to assure them that he would only amend it. I need not multiply instances, though they might be multiplied almost to a million. The note of the age is to suggest that the past may just as well be praised, since it cannot be mended. Men actually in that past have toiled like ants and died like locusts to undo some previous settlement that seemed secure; but we cannot do so much as repeal an Act of Parliament. We entertain the weak-minded notion that what is done can't be undone. Our view was well summarised in a typical Victorian song with the refrain: "The mill will never grind again the water that is past." There are many answers to this. One (which would involve a disquisition on the phenomena of Evaporation and Dew) we will here avoid. Another is, that to the minds of simple country folk, the object of a mill is not to grind water, but to grind corn, and that (strange as it may seem) there really have been societies sufficiently vigilant and valiant to prevent their corn perpetually flowing away from them, to the tune of a sentimental song.

Now this modern refusal to undo what has been done is not only an intellectual fault; it is a moral fault also. It is not merely our mental inability to understand the mistake we have made. It is also our spiritual refusal to admit that we have made a mistake. It was mere vanity in Mr. Brummell when he sent away trays full of imperfectly knotted neckcloths, lightly remarking, "These are our failures." It is a good instance of the nearness of vanity to humility, for at least he had to admit that they were failures. But it would have been spiritual pride in Mr. Brummell if he had tied on all the cravats, one on top of the other, lest his valet should discover that he had ever tied one badly. For in spiritual pride there is always an element of secrecy and solitude. Mr. Brummell would be satanic; also (which I fear would affect him more) he would be badly dressed. But he would be a perfect presentation of the modern publicist, who cannot do anything right, because he must not admit that he ever did anything wrong.

This strange, weak obstinacy, this persistence in the wrong path of progress, grows weaker and worse, as do all such weak things. And by the time in which I write its moral attitude has taken on something of the sinister and even the horrible. Our mistakes have become our secrets. Editors and journalists tear up with a guilty air all that reminds them of the party promises unfulfilled, or the party ideals reproaching them. It is true of our statesmen (much more than of our bishops, of whom Mr. Wells said it), that socially in evidence they are intellectually in hiding. The society is heavy with unconfessed sins; its mind is sore and silent with painful subjects; it has a constipation of conscience. There are many things it has done and allowed to be done which it does not really dare to think about; it calls them by other names and tries to talk itself into faith in a false past, as men make up the things they would have said in a quarrel. Of these sins one lies buried deepest but most noisome, and though it is stifled, stinks:

the true story of the relations of the rich man and the poor in England. The half-starved English proletarian is not only nearly a skeleton, but he is a skeleton in a cupboard.

## The True Story of Rich and Poor

It may be said, in some surprise, that surely we hear to-day on every side the same story of the destitute proletariat and the social problem, of the sweating in the unskilled trades or the overcrowding in the slums. It is granted; but I said the true story. Untrue stories there are in plenty, on all sides of the discussion. There is the interesting story of the Class Conscious Proletarian of All Lands, the chap who has "solidarity," and is always just going to abolish war. The Marxian Socialists will tell you all about him; only he isn't there. A common English workman is just as incapable of thinking of a German as anything but a German as he is of thinking of himself as anything but an Englishman. Then there is the opposite story; the story of the horrid man who is an atheist and wants to destroy the home, but who, for some private reason, prefers to call this Socialism. He isn't there either. The prosperous Socialists have homes exactly like yours and mine; and the poor Socialists are not allowed by the Individualists to have any at all. There is the story of the Two Workmen, which is a very nice and exciting story, about how one passed all the public houses in Cheapside and was made Lord Mayor on arriving at the Guildhall, while the other went into all the public houses and emerged quite ineligible for such a dignity. Alas! for this also is vanity. A thief might become Lord Mayor, but an honest workman certainly couldn't. Then there is the story of "The Relentless Doom," by which rich men were, by economic laws, forced to go on taking away money from poor men, although they simply longed to leave off: this is an unendurable thought to a free and Christian man, and the reader will be relieved to hear that it never happened. The rich could have left off stealing whenever they wanted to leave off, only this never happened either. Then there is the story of the cunning Fabian who sat on six committees at once and so coaxed the rich man to become quite poor. By simply repeating, in a whisper, that there are "wheels within wheels," this talented man managed to take away the millionaire's motor car, one wheel at a time, till the millionaire had quite forgotten that he ever had one. It was very clever of him to do this, only he has not done it. There is not a screw loose in the millionaire's motor, which is capable of running over the Fabian and leaving him a flat corpse in the road at a moment's notice. All these stories are very fascinating stories to be told by the Individualist and Socialist in turn to the great Sultan of Capitalism, because if they left off amusing him for an instant he would cut off their heads. But if they once began to tell the true story of the Sultan to the Sultan, he would boil them in oil; and this they wish to avoid.

The true story of the sin of the Sultan he is always trying, by listening to these stories, to forget. As we have said before in this chapter, he would prefer not to remember, because he has made up his mind not to repent. It is a curious story, and I shall try to tell it truly in the two chapters that follow. In all ages the tyrant is hard because he is soft. If his car crashes over bleeding and accusing crowds, it is because he has chosen the path of least resistance. It is because it is much easier to ride down a human race than ride up a moderately steep hill. The fight of the oppressor is always a pillow-fight; commonly a war with cushions—always a war for cushions. Saladin, the great Sultan, if I remember rightly, accounted it the greatest feat of swordsmanship to cut a cushion. And so indeed it is, as all of us can attest who have been for years past trying to cut into the swollen and windy corpulence of the modern compromise, that is at once cosy and cruel. For there is really in our world to-day the colour and silence of the cushioned divan; and that sense of palace within palace and garden within garden which makes the rich irresponsibility of the East. Have we not already the wordless dance, the wineless banquet, and all that strange unchristian conception of luxury without laughter? Are we not already in an evil Arabian Nights, and walking the nightmare cities of an invisible despot? Does not our hangman strangle secretly, the bearer of the bow string? Are we not already eugenists—that is, eunuch-makers? Do we not see the bright eyes, the motionless faces, and all that presence of something that is dead and yet sleepless? It is the presence of the sin that is sealed with pride and impenitence; the story of how the Sultan got his throne. But it is not the story he is listening to just now, but another story which has been invented to cover it—the story called "Eugenius: or the Adventures of One Not Born," a most varied and entrancing tale, which never fails to send him to sleep.

### Some Not Allowed to be Parents

The eugenist has every right to say, and must never cease saying, that many children are born who should never have been born, or, rather, who should never have been conceived. He has every right to say that the feeble-minded, and the alcoholic, and the insane, and those afflicted with venereal disease, must be so guarded and treated in future that they shall not become parents at all.—C.W. SALEEBY, 1914

# 10

# The History of a Tramp

There is one human thing left it is much harder to take from him. Debased by him and his betters, it is still something brought out of Eden, where God made him a demigod: it does not depend on money and but little on time. He can create in his own image. The terrible truth is in the heart of a hundred legends and mysteries. As Jupiter could be hidden from all-devouring Time, as the Christ Child could be hidden from Herod—so the child unborn is still hidden from the omniscient oppressor. He who lives not yet, he and he alone is left; and they seek his life to take it away.

He awoke in the Dark Ages and smelt dawn in the dark, and knew he was not wholly a slave. It was as if, in some tale of Hans Andersen, a stick or a stool had been left in the garden all night and had grown alive and struck root like a tree. For this is the truth behind the old legal fiction of the servile countries, that the slave is a "chattel," that is a piece of furniture like a stick or a stool. In the spiritual sense, I am certain it was never so unwholesome a fancy as the spawn of Nietzsche suppose to-day. No human being, pagan or Christian, I am certain, ever thought of another human being as a chair or a table. The mind cannot base itself on the idea that a comet is a cabbage; nor can it on the idea that a man is a stool. No man was ever unconscious of another's presence—or even indifferent to another's opinion. The lady who is said to have boasted her indifference to being naked before male slaves was showing off—or she meant something different. The lord who fed fishes by killing a slave was indulging in what most cannibals indulge in—a satanist affectation. The lady was consciously shameless and the lord was consciously cruel. But it simply is not in the human reason to carve men like wood or examine women like ivory, just as it is not in the human reason to think that two and two make five.

But there was this truth in the legal simile of furniture: that the slave, though certainly a man, was in one sense a dead man; in the sense that he was *moveable*. His locomotion was not his own: his master moved his arms and legs for him as if he were a marionette. Now it is important in the first degree to realise here what would be involved in such a fable as I have imagined, of a stool rooting itself like a shrub. For the general modern notion certainly is that life and liberty are in some way to be associated with novelty and not standing still. But it is just because the stool is lifeless that it moves about. It is just because the tree is alive that it does stand still. That was the main difference between the pagan slave and the Christian serf. The serf still belonged to the lord, as the stick that struck root in the garden would have still belonged to the owner of the garden; but it would have become a *live* possession. Therefore the owner is forced, by the laws of nature, to treat it with *some* respect; something becomes due from him. He cannot pull it up without killing it; it has gained a *place* in the garden—or the society. But the moderns are quite wrong in supposing that mere change and holiday and variety have necessarily any element of this life that is the only seed of liberty. You may say if you like that an employer, taking all his workpeople to a new factory in a Garden City, is giving them the greater freedom of forest landscapes and smokeless skies. If it comes to that, you can say that the slave-traders took negroes from their narrow and brutish African hamlets, and gave them the polish of foreign travel and medicinal breezes of a sea-voyage. But the tiny seed of citizenship and independence there already was in the serfdom of the Dark Ages, had nothing to do with what nice things the lord might do to the serf. It lay in the fact that there were some nasty things he could not do to the serf—there were not many, but there were some, and one of them was eviction. He could not make the serf utterly landless and desperate, utterly without access to the means of production, though doubtless it was rather the field that owned the serf, than the serf that owned the field. But even if you call the serf a beast of the field, he was not what we have tried to make the town workman—a beast with no field. Foulon said of the French peasants, "Let them eat grass." If he had said it of the modern London proletariat, they might well reply, "You have not left us even grass to eat."

There was, therefore, both in theory and practice, *some* security for the serf, because he had come to life and rooted. The seigneur could not wait in the field in all weathers with a battle-axe to prevent the serf scratching any living out of the ground, any more than the man in my fairy-tale could sit out in the garden all night with an umbrella to prevent the shrub getting any rain. The relation of lord and serf, therefore, involves a combination of two things: inequality and security. I know there are people who will at once point wildly to all sorts of examples, true and false, of insecurity of life in the Middle Ages; but these are people who do not grasp what we mean by the characteristic institutions of a society. For the matter of that, there are plenty of examples of equality in the Middle Ages, as the craftsmen in their guild or the monks electing their abbot. But just as modern England is not a feudal country, though there is a quaint survival called Heralds' College—or Ireland is not a commercial country, though there is a quaint survival called Belfast—it is true of the bulk and shape of that society that came out of the Dark Ages and ended at the Reformation, that it did not care about giving everybody an equal position, but did care about giving everybody a position. So that by the very beginning of that time even the slave had become a slave one could not get rid of, like the Scotch servant who stubbornly asserted that if his master didn't know a good servant he knew a good master. The free peasant, in ancient or modern times, is free to go or stay. The slave, in ancient times, was free neither to go nor stay. The serf was not free to go; but he was free to stay.

Now what have we done with this man? It is quite simple. There is no historical complexity about it in that respect.We have taken away his freedom to stay. We have turned him out of his field, and whether it was injustice, like turning a free farmer out of his field, or only cruelty to animals, like turning a cow out of its field, the fact remains that he is out in the road. First and last, we have simply destroyed the security, We have not in the least destroyed the inequality. All classes, all creatures, kind or cruel, still see this lowest stratum of society as separate from the upper strata and even the middle strata; he is as separate as the serf. A monster fallen from Mars, ignorant of our simplest word, would know the tramp was at the bottom of the ladder, as well as he would have known it of the serf. The walls of mud are no longer round his boundaries, but only round his boots. The coarse, bristling hedge is at the end of his chin, and not of his garden. But mud and bristles still stand out round him like a horrific halo, and separate him from his kind. The Martian would have no difficulty in seeing he was the poorest person in the nation. It is just as impossible that he should marry an heiress, or fight a duel with a duke, or contest a seat at Westminister or enter a club in Pall Mall, or take a scholarship at Balliol, or take a seat at an opera, or propose a good law, or protest against a bad one, as it was impossible to the serf. Where he differs is in something very different. He has lost what was possible to the serf. He can no longer scratch the bare earth by day or sleep on the bare earth by night, without being collared by a policeman.

Now when I say that this man has been oppressed as hardly any other man on this earth has been oppressed, I am not using rhetoric: I have a clear meaning which I am confident of explaining to any honest reader. I do not say he has been treated worse: I say he has been treated differently from the unfortunate in all ages. And the difference is this: that all the others were told to do something, and killed or tortured if they did anything else. This man is not told to do something: he is merely forbidden to do anything. When he was a slave, they said to him, "Sleep in this shed; I will beat you if you sleep anywhere else." When he was a serf, they said to him, "Let me find you in this field: I will hang you if I find you in anyone else's field." But now he is a tramp they say to him, "You shall be jailed if I find you in anyone else's field: *but I will not give you a field.*" They say, "You shall be punished if you are caught sleeping outside your shed: *but there is no shed.*" If you say that modern magistracies could never say such mad contradictions, I answer with entire certainty that they do say them. A little while ago two tramps were summoned before a magistrate, charged with sleeping in the open air when they had nowhere else to sleep. But this is not the full fun of the incident. The real fun is that each of them eagerly produced about twopence, to prove that they could have got a bed, but deliberately didn't. To which the policeman replied that twopence would not have got them a bed: that they could not possibly have got a bed: and *therefore* (argued that thoughtful officer) they ought to be punished for not getting one. The intelligent magistrate was much struck with the argument: and proceeded to imprison these two men for not doing a thing they could not do. But he was careful to explain that if they had sinned needlessly and in wanton lawlessness, they would have left the court without a stain on their characters; but as they could not avoid it, they were very much to blame. These things are being done in every part of England every day. They have their parallels even in every daily paper; but they have no parallel in any other earthly people or period; except in that insane command to make bricks without straw which brought down all the plagues of Egypt. For the common historical joke about Henry VIII hanging a man for being Catholic and burning him for being Protestant is a symbolic joke only. The sceptic in the Tudor time could do something: he could always agree with Henry VIII. The desperate man to-day can do nothing. For you cannot agree with a maniac who sits on the bench with the straws sticking out of his hair and says, "Procure threepence from nowhere and I will give you leave to do without it."

If it be answered that he can go to the workhouse, I reply that such an answer is founded on confused thinking. It is true that he is free to go to the workhouse, but only in the same sense in which he is free to go to jail, only in the same sense in which the serf under the gibbet was free to find peace in the grave. Many of the poor greatly prefer the grave to the workhouse, but that is not at all my argument here. The point is this: that it could not have been the general policy of a lord towards serfs to kill them all like wasps. It could not have been his standing "Advice to Serfs" to say, "Get hanged." It cannot be the standing advice of magistrates to citizens to go to prison. And, precisely as plainly, it cannot be the standing advice of rich men to very poor men to go to the workhouses. For that would mean the rich raising their own poor rates enormously to keep a vast and expensive establishment of slaves. Now it may come to this, as Mr. Belloc maintains, but it is not the theory on which what we call the workhouse does in fact rest. The very shape (and even the very size) of a workhouse express the fact that it was founded for certain quite exceptional human failures—like the lunatic asylum. Say to a man, "Go to the madhouse," and he will say, "Wherein am I mad?" Say to a tramp under a hedge, "Go to the house of exceptional failures," and he will say with equal reason, "I travel because I have no house; I walk because I have no horse; I sleep out because I have no bed. Wherein have I failed?" And he may have the intelligence to add, "Indeed, your worship, if somebody has failed, I think it is not I." I concede, with all due haste, that he might perhaps say "me."

The speciality then of this man's wrong is that it is the only historic wrong that has in it the quality of *nonsense*. It could only happen in a nightmare; not in a clear and rational hell. It is the top point of that anarchy in the governing mind which, as I said at the beginning, is the main trait of modernity, especially in England. But if the first note in our policy is madness, the next note is certainly meanness. There are two peculiarly mean and unmanly legal mantraps in which this wretched man is tripped up. The first is that which prevents him from doing what any ordinary savage or nomad would do—take his chance of an uneven subsistence on the rude bounty of nature.

There is something very abject about forbidding this; because it is precisely this adventurous and vagabond spirit which the educated classes praise most in their books, poems and speeches. To feel the drag of the roads, to hunt in nameless hills and fish in secret streams, to have no address save "Over the Hills and Far Away," to be ready to breakfast on berries and the daybreak and sup on the sunset and a sodden crust, to feed on wild things and be a boy again, all this is the heartiest and sincerest impulse in recent culture, in the songs and tales of Stevenson, in the cult of George Borrow and in the delightful little books published by Mr. E. V. Lucas. It is the one true excuse in the core of Imperialism; and it faintly softens the squalid prose and wooden-headed wickedness of the Self-Made Man who "came up to London with twopence in his pocket." But when a poorer but braver man with less than twopence in his pocket does the very thing we are always praising, makes the blue. heavens his house, we send him to a house built for infamy and flogging. We take poverty itself and only permit it with a property qualification; we only allow a man to be poor if he is rich. And we do this most savagely if he has sought to snatch his life by that particular thing of which our boyish adventure stories are fullest—hunting and fishing. The extremely severe English game laws hit most heavily what the highly reckless English romances praise most irresponsibly. All our literature is full of praise of the chase—especially of the wild goose chase. But if a poor man followed, as Tennyson says, "far as the wild swan wings to where the world dips down to sea and sands," Tennyson would scarcely allow him to catch it. If he found the wildest goose in the wildest fenland in the wildest regions of the sunset, he would very probably discover that the rich never sleep; and that there are no wild things in England.

In short, the English ruler is always appealing to a nation of sportsmen and concentrating all his efforts on preventing them from having any sport. The Imperialist is always pointing out with exultation that the common Englishman can live by adventure anywhere on the globe, but if the common Englishman tries to live by adventure in England, he is treated as harshly as a thief, and almost as harshly as an honest journalist. This is hypocrisy: the magistrate who gives his son *Treasure Island* and then imprisons a tramp is a hypocrite; the squire who is proud of English colonists and indulgent to English schoolboys, but cruel to English poachers, is drawing near that deep place wherein all liars have their part. But our point here is that the baseness is in the idea of *bewildering* the tramp; of leaving him no place for repentance. It is quite true, of course, that in the days of slavery or of serfdom the needy were fenced by yet fiercer penalties from spoiling the hunting of the rich. But in the older case there were two very important differences, the second of which is our main subject in this chapter. The first is that in a comparatively wild society, however fond of hunting, it seems impossible that enclosing and game-keeping can have been so omnipresent and efficient as in a society full of maps and policemen. The second difference is the one already noted: that if the slave or semi-slave was forbidden to get his food in the greenwood, he was told to get it somewhere else. The note of unreason was absent.

This is the first meanness; and the second is like unto it. If there is one thing of which cultivated modern letters is full besides adventure it is altruism. We are always being told to help others, to regard our wealth as theirs, to do what good we can, for we shall not pass this way again. We are everywhere urged by humanitarians to help lame dogs over stiles—though some humanitarians, it is true, seem to feel a colder interest in the case of lame men and women. Still, the chief fact of our literature, among all historic literatures, is human charity. But what is the chief fact of our legislation? The great outstanding fact of modern legislation, among all historic legislations, is the forbidding of human charity. It is this astonishing paradox, a thing in the teeth of all logic and conscience, that a man that takes another man's money with his leave can be punished as if he had taken it without his leave. All through those dark or dim ages behind us, through times of servile stagnation, of feudal insolence, of pestilence and civil strife and all else that can war down the weak, for the weak to ask for charity was counted lawful, and to give that charity, admirable. In all other centuries, in short, the casual bad deeds of bad men could be partly patched and mended by the casual good deeds of good men. But this is now forbidden; for it would leave the tramp a last chance if he could beg.

Now it will be evident by this time that the interesting scientific experiment on the tramp entirely depends on leaving him *no* chance, and not (like the slave) one chance. Of the economic excuses offered for the persecution of beggars it will be more natural to speak in the next chapter. It will suffice here to say that they are mere excuses, for a policy that has been persistent while probably largely unconscious, with a selfish and atheistic unconsciousness. That policy was directed towards something—or it could never have cut so cleanly and cruelly across the sentimental but sincere modern trends to adventure and altruism. Its object is soon stated. It was directed towards making the very poor man work for the capitalist, for any wages or none. But all this, which I shall also deal with in the next chapter, is here only important as introducing the last truth touching the man of despair. The game laws have taken from him his human command of Nature. The mendicancy laws have taken from him his human demand on Man. There is one human thing left it is much harder to take from him. Debased by him and his betters, it is still something brought out of Eden, where God made him a demigod: it does not depend on money and but little on time. He can create in his own image. The terrible truth is in the heart of a hundred legends and mysteries. As Jupiter could be hidden from all-devouring Time, as the Christ Child could be hidden from Herod—so the child unborn is still hidden from the omniscient oppressor. He who lives not yet, he and he alone is left; and they seek his life to take it away.

# 11

# The History of a Eugenist

This period can be dated practically by the period when the old and genuine Protestant religion of England began to fail; and the average business man began to be agnostic, not so much because he did not know where he was, as because he wanted to forget. Many of the rich took to scepticism exactly as the poor took to drink; because it was a way out. But in any case, the man who had made a mistake not only refused to unmake it, but decided to go on making it. But in this he made yet another most amusing mistake, which was the beginning of all Eugenics.

He does not live in a dark lonely tower by the sea, from which are heard the screams of vivisected men and women. On the contrary, he lives in Mayfair. He does not wear great goblin spectacles that magnify his eyes to moons or diminish his neighbours to beetles. When he is more dignified he wears a single eyeglass; when more intelligent, a wink. He is not indeed wholly without interest in heredity and Eugenical biology; but his studies and experiments in this science have specialised almost exclusively in *equus celer,* the rapid or running horse. He is not a doctor; though he employs doctors to work up a case for Eugenics, just as he employs doctors to correct the errors of his dinner. He is not a lawyer, though unfortunately often a magistrate. He is not an author or a journalist; though he not infrequently owns a newspaper. He is not a soldier, though he may have a commission in the yeomanry; nor is he generally a gentleman, though often a nobleman. His wealth now commonly comes from a large staff of employed persons who scurry about in big buildings while he is playing golf. But he very often laid the foundations of his fortune in a very curious and poetical way, the nature of which I have never fully understood. It consisted in his walking about the street without a hat and going up to another man and saying, "Suppose I have two hundred whales out of the North Sea." To which the other man replied, "And let us imagine that I am in possession of two thousand elephants' tusks." They then exchange, and the first man goes up to a third man and says, "Supposing me to have lately come into the possession of two thousand elephants' tusks, would you, etc.? If you play this game well, you become very rich; if you play it badly you have to kill yourself or try your luck at the Bar. The man I am speaking about must have played it well, or at any rate successfully.

He was born about 1860; and has been a member of Parliament since about 1890. For the first half of his life he was a Liberal; for the second half he has been a Conservative; but his actual policy in Parliament has remained largely unchanged and consistent. His policy in Parliament is as follows: he takes a seat in a room downstairs at Westminster, and takes from his breast pocket an excellent cigar-case, from which in turn he takes an excellent cigar. This he lights, and converses with other owners of such cigars on *equus celer* or such matters as may afford him entertainment. Two or three times in the afternoon a bell rings; whereupon he deposits the cigar in an ashtray with great particularity, taking care not to break the ash, and proceeds to an upstairs room, flanked with two passages. He then walks into whichever of the two passages shall be indicated to him by a young man of the upper classes, holding a slip of paper. Having gone into this passage he comes out of it again, is counted by the young man and proceeds downstairs again; where he takes up the cigar once more, being careful not to break the ash. This process, which is known as Representative Government, has never called for any great variety in the manner of his life. Nevertheless, while his Parliamentary policy is unchanged, his change from one side of the House to the other did correspond with a certain change in his general policy in commerce and social life. The change of the party label is by this time quite a trifling matter; but there was in his case a change of philosophy or at least a change of project; though it was not so much becoming a Tory, as becoming rather the wrong kind of Socialist. He is a man with a history. It is a sad history, for he is certainly a less good man than he was when he started. That is why he is the man who is really behind Eugenics. It is because he has degenerated that he has come to talking of Degeneration.

In his Radical days (to quote from one who corresponded in some ways to this type) he was a much better man, because he was a much less enlightened one. The hard impudence of his first Manchester Individualism was softened by two relatively humane qualities; the first was a much greater manliness in his pride the second was a much greater sincerity in his optimism. For the first point, the modern capitalist is merely industrial; but this man was also industrious. He was proud of hard work; nay, he was even proud of low work—if he could speak of it in the past and not the present. In fact, he invented a new kind of Victorian snobbishness, an inverted snobbishness. While the snobs of Thackeray turned Muggins into De Mogyns, while the snobs of Dickens wrote letters describing themselves as officers' daughters "accustomed to every luxury—except spelling," the Individualist spent his life in hiding his prosperous parents. He was more like an American plutocrat when he began; but he has since lost the American simplicity. The Frenchman works until he can play. The American works until he can't play; and then thanks the devil, his master, that he is donkey enough to die in harness. But the Englishman, as he has since become, works until he can pretend that he never worked at all. He becomes as far as possible another person—a country gentleman who has never heard of his shop; one whose left hand holding a gun knows not what his right hand doeth in a ledger. He uses a peerage as an alias, and a large estate as a sort of alibi. A stern Scotch minister remarked concerning the game of golf, with a terrible solemnity of manner, "the man who plays golf—he neglects his business, he forsakes his wife, he forgets his God." He did not seem to realise that it is the chief aim of many a modern capitalist's life to forget all three.

This abandonment of a boyish vanity in work, this substitution of a senile vanity in indolence, this is the first respect in which the rich Englishman has fallen. He was more of a man when he was at least a master workman and not merely a master. And the second important respect in which he was better at the beginning is this: that he did then, in some hazy way, half believe that he was enriching other people as well as himself. The optimism of the early Victorian Individualists was not wholly hypocritical. Some of the clearest-headed and blackest-hearted of them, such as Malthus, saw where things were going, and boldly based their Manchester city on pessimism instead of optimism. But this was not the general case; most of the decent rich of the Bright and Cobden sort did have a kind of confused faith that the economic conflict would work well in the long run for everybody. They thought the troubles of the poor were incurable by State action (they thought that of all troubles), but they did not cold-bloodedly contemplate the prospect of those troubles growing worse and worse. By one of those tricks or illusions of the brain to which the luxurious are subject in all ages, they sometimes seemed to feel as if the populace had triumphed symbolically in their own persons. They blasphemously thought about their thrones of gold what can only be said about a cross—that they, being lifted up, would draw all men after them. They were so full of the romance that anybody could be Lord Mayor, that they seemed to have slipped into thinking that everybody could. It seemed as if a hundred Dick Whittingtons, accompanied by a hundred cats, could all be accommodated at the Mansion House. It was all nonsense but it was not (until later) all humbug.

## The Horrible Discovery

Step by step, however, with a horrid and increasing clearness, this man discovered what he was doing. It is generally one of the worst discoveries a man can make. At the beginning, the British plutocrat was probably quite as honest in suggesting that every tramp carried a magic cat like Dick Whittington, as the Bonapartist patriot was in saying that every French soldier carried a marshal's *baton* in his knapsack. But it is exactly here that the difference and the danger appears. There is no comparison between a well-managed thing like Napoleon's army and an unmanageable thing like modern competition. Logically, doubtless, it was impossible that every soldier should carry a marshal's *baton;* they could not all be marshals any more than they could all be mayors. But if the French soldier did not always have a *baton* in his knapsack, he always had a knapsack. But when that Self-Helper who bore the adorable name of Smiles told the English tramp that he carried a coronet in his bundle, the English tramp had an unanswerable answer. He pointed out that he had no bundle. The powers that ruled him had not fitted him with a knapsack, any more than they had fitted him with a future—or even a present. The destitute Englishman, so far from hoping to become anything, had never been allowed even to be anything. The French soldier's ambition may have been in practice not only a short, but even a deliberately shortened ladder, in which the top rungs were knocked out. But for the English it was the bottom rungs that were knocked out, so that they could not even begin to climb. And sooner or later, in exact proportion to his intelligence, the English plutocrat began to understand not only that the poor were impotent, but that their impotence had been his only power. The truth was not merely that his riches had left them poor; it was that nothing but their poverty could have been strong enough to make him rich. It is this paradox, as we shall see, that creates the curious difference between him and every other kind of robber.

I think it is no more than justice to him to say that the knowledge, where it has come to him, has come to him slowly; and I think it came (as most things of common sense come) rather vaguely and as in a vision— that is, by the mere look of things. The old Cobdenite employer was quite within his rights in arguing that earth is not heaven, that the best obtainable arrangement might contain many necessary evils; and that Liverpool and Belfast might be growing more prosperous as a whole in spite of pathetic things that might be seen there. But I simply do not believe he has been able to look at Liverpool and Belfast and continue to think this: that is why he has turned himself into a sham country gentleman. Earth is not heaven, but the nearest we can get to heaven ought not to *look* like hell; and Liverpool and Belfast look like hell, whether they are or not. Such cities might be growing prosperous as a whole, though a few citizens were more miserable. But it was more and more broadly apparent that it was exactly and precisely *as a whole* that they were not growing more prosperous, but only the few citizens who were growing more prosperous by their increasing misery. You could not say a country was becoming a white man's country when there were more and more black men in it every day. You could not say a community was more and more masculine when it was producing more and more women. Nor can you say that a city is growing richer and richer when more and more of its inhabitants are very poor men. There might be a false agitation founded on the pathos of individual cases in a community pretty normal in bulk. But the fact is that no one can take a cab across Liverpool without having a quite complete and unified impression that the pathos is not a pathos of individual cases, but a pathos in bulk. People talk of the Celtic sadness; but there are very few things in Ireland that look so sad as the Irishman in Liverpool. The desolation of Tara is cheery compared with the desolation of Belfast. I recommend Mr. Yeats and his mournful friends to turn their attention to the pathos of Belfast. I think if they hung up the harp that [was] once in Lord Furness's factory, there would be a chance of another string breaking.

Broadly, and as things bulk to the eye, towns like Leeds, if placed beside towns like Rouen or Florence, or Chartres, or Cologne, do actually look like beggars walking among burghers. After that overpowering and unpleasant impression it is really useless to argue that they are richer because a few of their parasites get rich enough to live somewhere else. The point may be put another way, thus: that it is not so much that these more modern cities have this or that monopoly of good or evil; it is that they have every good in its fourth-rate form and every evil in its worst form. For instance, that interesting weekly paper *The Nation* amiably rebuked Mr. Belloc and myself for suggesting that revelry and the praise of fermented liquor were more characteristic of Continental and Catholic communities than of communities with the religion and civilisation of Belfast. It said that if we would "cross the border" into Scotland, we should find out our mistake. Now, not only have I crossed the border, but

I have had considerable difficulty in crossing the road in a Scotch town on a festive evening. Men were literally lying like piled-up corpses in the gutters, and from broken bottles whisky was pouring down the drains. I am not likely, therefore, to attribute a total and arid abstinence to the whole of industrial Scotland. But I never said that drinking was a mark rather of the Catholic countries. I said that *moderate* drinking was a mark rather of the Catholic countries. In other words, I say of the common type of Continental citizen, not that he is the only person who is drinking, but that he is the only person who knows how to drink. Doubtless gin is as much a feature of Hoxton as beer is a feature of Munich. But who is the connoisseur who prefers the gin of Hoxton to the beer of Munich? Doubtless the Protestant Scotch ask for "Scotch," as the men of Burgundy ask for Burgundy. But do we find them lying in heaps on each side of the road when we walk through a Burgundian village? Do we find the French peasant ready to let Burgundy escape down a drain-pipe? Now this one point, on which I accept *The Nation's* challenge, can be exactly paralleled on almost every point by which we test a civilisation. It does not matter whether we are for alcohol or against it. On either argument Glasgow is more objectionable than Rouen. The French abstainer makes less fuss; the French drinker gives less offence. It is so with property, with war, with everything. I can understand a teetotaler being horrified, on his principles, at Italian wine-drinking. I simply cannot believe he could be *more* horrified at it than at Hoxton gin-drinking. I can understand a Pacifist, with his special scruples, disliking the militarism of Belfort. I flatly deny that he can dislike it *more* than the militarism of Berlin. I can understand a good Socialist hating the petty cares of the distributed peasant property. I deny that any good Socialist can hate them *more* than he hates the large cares of Rockefeller. That is the unique tragedy of the plutocratic state to-day; it has *no* successes to hold up against the failures it alleges to exist in Latin or other methods. You can (if you are well out of his reach) call the Irish rustic debased and superstitious. I defy you to contrast his debasement and superstition with the citizenship and enlightenment of the English rustic.

## Knowing He is a Cancer

To-day the rich man knows in his heart that he is a cancer and not an organ of the State. He differs from all other thieves or parasites for this reason: that the brigand who takes by force wishes his victims to be rich. But he who wins by a one-sided contract actually wishes them to be poor. Rob Roy in a cavern, hearing a company approaching, will hope (or if in a pious mood, pray) that they may come laden with gold or goods. But Mr.

Rockefeller, in his factory, knows that if those who pass are laden with goods they will pass on. He will therefore (if in a pious mood) pray that they may be destitute, and so be forced to work his factory for him for a starvation wage. It is said (and also, I believe, disputed) that Blücher riding through the richer parts of London exclaimed, "What a city to sack!" But Blücher was a soldier if he was a bandit. The true sweater feels quite otherwise. It is when he drives through the poorest parts of London that he finds the streets paved with gold, being paved with prostrate servants; it is when he sees the grey lean leagues of Bow and Poplar that his soul is uplifted and he knows he is secure. This is not rhetoric, but economics.

I repeat that up to a point the profiteer was innocent because he was ignorant; he had been lured on by easy and accommodating events. He was innocent as the new Thane of Glamis was innocent, as the new Thane of Cawdor was innocent; but the King—The modern manufacturer, like Macbeth, decided to march on, under the mute menace of the heavens. He knew that the spoil of the poor was in his houses; but he could not, after careful calculation, think of any way in which they could get it out of his houses without being arrested for housebreaking. He faced the future with a face flinty with pride and impenitence. This period can be dated practically by the period when the old and genuine Protestant religion of England began to fail; and the average business man began to be agnostic, not so much because he did not know where he was, as because he wanted to forget. Many of the rich took to scepticism exactly as the poor took to drink; because it was a way out. But in any case, the man who had made a mistake not only refused to unmake it, but decided to go on making it. But in this he made yet another most amusing mistake, which was the beginning of all Eugenics.

### Noted Leaders Support Controlled Breeding

Chicago.—Selective immigration, sterilization of defectives and control of everything having to do with the reproduction of human beings are among the objects of the Eugenics Committee of the United States, which has just issued its sweeping programme for the betterment of the human race.

The committee comes out frankly for birth control and the most widespread distribution of knowledge concerning the breeding of the species.

President Emeritus Charles W. Eliot of Harvard; Senator Royal S. Copeland, former Health Commissioner of New York; Surgeon General H. S. Cummings, Washington; President Livingston Farrand of Cornell University; Dr. David Starr Jordan, Dr. Ray Liman Wilbur, Dr. Charles E. Sawyer and many other noted educational, medical, and social welfare leaders are on the committee's advisory council.—BIRTH CONTROL NEWS, 1923

# 12

# The Vengeance of the Flesh

So at least it seemed, doubtless in a great degree subconsciously, to the man who had wagered all his wealth on the usefulness of the poor to the rich and the dependence of the rich on the poor. The time came at last when the rather reckless breeding in the abyss below ceased to be a supply, and began to be something like a wastage; ceased to be something like keeping foxhounds, and began alarmingly to resemble a necessity of shooting foxes.

By a quaint paradox, we generally miss the meaning of simple stories because we are not subtle enough to understand their simplicity. As long as men were in sympathy with some particular religion or other romance of things in general, they saw the thing solid and swallowed it whole, knowing that it could not disagree with them. But the moment men have lost the instinct of being simple in order to understand it, they have to be very subtle in order to understand it. We can find, for instance, a very good working case in those old puritanical nursery tales about the terrible punishment of trivial sins; about how Tommy was drowned for fishing on the Sabbath, or Sammy struck by lightning for going out after dark. Now these moral stories are immoral, because Calvinism is immoral. They are wrong, because Puritanism is wrong. But they are not quite so wrong, they are not a quarter so wrong, as many superficial sages have supposed.

The truth is that everything that ever came out of a human mouth had a human meaning; and not one of the fixed fools of history was such a fool as he looks. And when our great-uncles or great-grandmothers told a child he might be drowned by breaking the Sabbath, their souls (though undoubtedly, as Touchstone said, in a parlous state) were not in quite so simple a state as is suggested by supposing that their god was a devil who

dropped babies into the Thames for a trifle. This form of religious literature is a morbid form if taken by itself; but it did correspond to a certain
reality in psychology which most people of any religion, or even of none,
have felt a touch of at some time or other. Leaving out theological terms
as far as possible, it is the subconscious feeling that one can be wrong
with Nature as well as right with Nature; that the point of wrongness may
be a detail (in the superstitions of heathens this is often quite a triviality);
but that if one is really wrong with Nature, there is no particular reason
why all her rivers should not drown or all her storm-bolts strike one who
is, by this vague yet vivid hypothesis, her enemy. This may be a mental
sickness, but it is too human or too mortal a sickness to be called solely a
superstition. It is not solely a superstition; it is not simply superimposed
upon human nature by something that has got on top of it. It flourishes
without check among non-Christian systems, and it flourishes especially
in Calvinism, because Calvinism is the most non-Christian of Christian
systems. But like everything else that inheres in the natural senses and
spirit of man, it has something in it; it is not stark unreason. If it is an ill
(and it generally is), it is one of the ills that flesh is heir to, but he is the
lawful heir. And like many other dubious or dangerous human instincts or
appetites, it is sometimes useful as a warning against worse things.

Now the trouble of the nineteenth century very largely came from the
loss of this; the loss of what we may call the natural and heathen mysti-
cism. When modern critics say that Julius Caesar did not believe in Jupiter, or that Pope Leo did not believe in Catholicism, they overlook an
essential difference between those ages and ours. Perhaps Julius did not
believe in Jupiter; but he did not disbelieve in Jupiter. There was nothing
in his philosophy, or the philosophy of that age, that could forbid him to
think that there was a spirit personal and predominant in the world. But
the modern materialists are not permitted to doubt; they are forbidden to
believe. Hence, while the heathen might avail himself of accidental
omens, queer coincidences or casual dreams, without knowing for certain
whether they were really hints from heaven or premonitory movements in
his own brain, the modern Christian turned heathen must not entertain
such notions at all, but must reject the oracle as the altar. The modern
sceptic was drugged against all that was natural in the supernatural. And
this was why the modern tyrant marched upon his doom, as a tyrant literally pagan might possibly not have done.

There is one idea of this kind that runs through most popular tales
(those, for instance, on which Shakespeare is so often based)—an idea
that is profoundly moral even if the tales are immoral. It is what may be
called the flaw in the deed: the idea that, if I take my advantage to the full,

I shall hear of something to my disadvantage. Thus Midas fell into a fallacy about the currency; and soon had reason to become something more than a Bimetallist. Thus Macbeth had a fallacy about forestry; he could not see the trees for the wood. He forgot that, though a place cannot be moved, the trees that grow on it can. Thus Shylock had a fallacy of physiology; he forgot that, if you break into the house of life, you find it a bloody house in the most emphatic sense. But the modern capitalist did not read fairytales, and never looked for the little omens at the turnings of the road. He (or the most intelligent section of him) had by now realised his position, and knew in his heart it was a false position. He thought a margin of men out of work was good for his business; he could no longer really think it was good for his country. He could no longer be the old "hardheaded" man who simply did not understand things; he could only be the hard-hearted man who faced them. But he still marched on; he was sure he had made no mistake.

However, he had made a mistake—as definite as a mistake in multiplication. It may be summarised thus: that the same inequality and insecurity that makes cheap labour may make bad labour, and at last no labour at all. It was as if a man who wanted something from an enemy, should at last reduce the enemy to come knocking at his door in the despair of winter, should keep him waiting in the snow to sharpen the bargain; and then come out to find the man dead upon the doorstep.

He had discovered the divine boomerang; his sin had found him out. The experiment of Individualism—the keeping of the worker half in and half out of work—was far too ingenious not to contain a flaw. It was too delicate a balance to work entirely with the strength of the starved and the vigilance of the benighted. It was too desperate a course to rely wholly on desperation. And as time went on the terrible truth slowly declared itself; the degraded class was really degenerating. It was right and proper enough to use a man as a tool; but the tool, ceaselessly used, was being used up. It was quite reasonable and respectable, of course, to fling a man away like a tool; but when it was flung away in the rain the tool rusted. But the comparison to a tool was insufficient for an awful reason that had already begun to dawn upon the master's mind. If you pick up a hammer, you do not find a whole family of nails clinging to it. If you fling away a chisel by the roadside, it does not litter and leave a lot of little chisels. But the meanest of the tools, Man, had still this strange privilege which God had given him, doubtless by mistake. Despite all improvements in machinery, the most important part of the machinery (the fittings technically described in the trade as "hands") were apparently growing worse. The firm was not only encumbered with one useless servant, but he immediately turned himself into five useless servants. "The poor should not be emancipated," the old reactionaries used to say, "until they are fit for freedom." But if this downrush went on, it looked as if the poor would not stand high enough to be fit for slavery.

So at least it seemed, doubtless in a great degree subconsciously, to the man who had wagered all his wealth on the usefulness of the poor to the rich and the dependence of the rich on the poor. The time came at last when the rather reckless breeding in the abyss below ceased to be a supply, and began to be something like a wastage; ceased to be something like keeping foxhounds, and began alarmingly to resemble a necessity of shooting foxes. The situation was aggravated by the fact that these sexual pleasures were often the only ones the very poor could obtain, and were, therefore, disproportionately pursued, and by the fact that their conditions were often such that prenatal nourishment and such things were utterly abnormal. The consequences began to appear. To a much less extent than the Eugenists assert, but still to a notable extent, in a much looser sense than the Eugenists assume, but still in some sort of sense, the types that were inadequate or incalculable or uncontrollable began to increase. Under the hedges of the country, on the seats of the parks, loafing under the bridges or leaning over the Embankment, began to appear a new race of men—men who are certainly not mad, whom we shall gain no scientific light by calling feeble-minded, but who are, in varying individual degrees, dazed or drink-sodden, or lazy or tricky or tired in body and spirit. In a far less degree than the teetotallers tell us, but still in a large degree, the traffic in gin and bad beer (itself a capitalist enterprise) fostered the evil, though it had not begun it. Men who had no human bond with the instructed man, men who seemed to him monsters and creatures without mind, became an eyesore in the market-place and a terror on the empty roads. The rich were afraid.

Moreover, as I have hinted before, the act of keeping the destitute out of public life, and crushing them under confused laws, had an effect on their intelligences which paralyses them even as a proletariat. Modern people talk of "Reason versus Authority"; but authority itself involves reason, or its orders would not even be understood. If you say to your valet, "Look after the buttons on my waistcoat," he may do it, even if you throw a boot at his head. But if you say to him, "Look after the buttons on my top-hat," he will not do it, though you empty a boot-shop over him. If you say to a schoolboy, "Write out that Ode of Horace from memory in the original Latin," he may do it without a flogging. If you say, "Write out that Ode of Horace in the original German," he will not do it with a thousand floggings. If you will not learn logic, he certainly will not learn

Latin. And the ludicrous laws to which the needy are subject (such as that which punishes the homeless for not going home) have really, I think, a great deal to do with a certain increase in their sheepishness and short-wittedness, and, therefore, in their industrial inefficiency. By one of the monstrosities of the feeble-minded theory, a man actually acquitted by judge and jury could *then* be examined by doctors as to the state of his mind—presumably in order to discover by what diseased eccentricity he had refrained from the crime. In other words, when the police cannot jail a man who is innocent of doing something, they jail him for being too innocent to do anything. I do not suppose the man is an idiot at all, but I can believe he feels more like one after the legal process than before. Thus all the factors—the bodily exhaustion, the harassing fear of hunger, the reckless refuge in sexuality, and the black botheration of bad laws—combined to make the employee more unemployable.

### Unemployed as Defective

The first category of defectives subject to be dealt with, viz., those who are found neglected, abandoned, or cruelly treated, has been extended to include those who are "without visible means of support"; an obvious improvement from all standpoints.—EUGENICS REVIEW, 1913

Now, it is very important to understand here that there were two courses of action still open to the disappointed capitalist confronted by the new peril of this real or alleged decay. First, he might have reversed his machine, so to speak, and started unwinding the long rope of dependence by which he had originally dragged the proletarian to his feet. In other words, he might have seen that the workmen had more money, more leisure, more luxuries, more status in the community, and then trusted to the normal instincts of reasonably happy human beings to produce a generation better born, bred and cared for than these tortured types that were less and less use to him. It might still not be too late to rebuild the human house upon such an architectural plan that poverty might fly out of the window, with the reasonable prospect of love coming in at the door. In short, he might have let the English poor, the mass of whom were not weak-minded, though more of them were growing weaker, a reasonable chance, in the form of more money, of achieving their eugenical resurrection themselves. It has never been shown, and it cannot be shown, that the method would have failed. But it can be shown, and it must be closely and clearly noted, that the method had very strict limitations from the employers' own point of view. If they made the worker too comfortable, he would not work to increase another's comforts; if they made him too independent, he would not work like a dependent. If, for instance, his wages were so good that he could save out of them, he might cease to be a wage-earner. If his house or garden were his own, he might stand an economic siege in it. The whole capitalist experiment had been built on his dependence; but now it was getting out of hand, not in the direction of freedom, but of frank helplessness. One might say that his dependence had got independent of control.

But there was another way. And towards this the employer's ideas began, first darkly and unconsciously, but now more and more clearly, to drift. Giving property, giving leisure, giving status costs money. But there is one human force that costs nothing. As it does not cost the beggar a penny to indulge, so it would not cost the employer a penny to employ. He could not alter or improve the tables or the chairs on the cheap. But there were two pieces of furniture (labelled respectively "the husband" and "the wife") whose relations were much cheaper. He could alter the *marriage* in the house in such a way as to promise himself the largest possible number of the kind of children he did want, with the smallest possible number of the kind he did not. He could divert the force of sex from producing vagabonds. And he could harness to his high engines unbought the red unbroken river of the blood of a man in his youth, as he has already harnessed to them all the wild waste rivers of the world.

### Marriage Bans

An influential correspondent, who has given much thought to the subject, but who desires to remain anonymous, sends the following brief outline of what he regards as an effective scheme of Eugenic reform. He is, we gather, driven to this conclusion because he holds that sterilization is impractical and the advocacy of birth control will relatively diminish the numbers of the fit. Coming from the quarter it does, the scheme is well worthy of consideration. We certainly earnestly wish that such an examination as is proposed could be brought within the region of practical politics.

1.—Everyone to undergo a medical and psychological examination as for Life Insurance at say 18, and to be classed under A, B, or C.

A.—First Class.

B.—Bodily and Mental condition good enough for mating.

C.—Sub-normals, who should not breed.

II.—No one to marry without making known his or her classification—to those who should know it.

III.—Sub-normals not to marry unless the woman is over 45 years of age.

There remains the case of illegitimate connexions. It might be made penal for a sub-normal to form such a connexion with a woman under 45.

—*EUGENICS REVIEW*, APR. 1923

# CHAPTER

# 13

# The Meanness of the Motive

The curious point is that the hopeful one concludes by saying, "When people have large families and small wages, not only is there a high infantile death-rate, but often those who do live to grow up are stunted and weakened by having had to share the family income for a time with those who died early. There would be less unhappiness if there were no unwanted children." You will observe that he tacitly takes it for granted that the small wages and the income, desperately shared, are the fixed points, like day and night, the conditions of human life. Compared with them marriage and maternity are luxuries, things to be modified to suit the wage-market.

Now, if any ask whether it be imaginable that an ordinary man of the wealthier type should analyse the problem or conceive the plan, the inhumanly farseeing plan, as I have set it forth, the answer is: "Certainly not." Many rich employers are too generous to do such a thing; many are too stupid to know what they are doing. The eugenical opportunity I have described is but an ultimate analysis of a whole drift of thoughts in the type of man who does not analyse his thoughts. He sees a slouching tramp, with a sick wife and a string of rickety children, and honestly wonders what he can do with them. But prosperity does not favour self-examination; and he does not even ask himself whether he means "How can I help them?" or "How can I use them?"—what he can still do for them, or what they could still do for him. Probably he sincerely means both, but the latter much more than the former; he laments the breaking of the tools of Mammon much more than the breaking of the images of God. It would be almost impossible to grope in the limbo of what he does think; but we can assert that there is one thing he doesn't think. He doesn't think, "This man might be as jolly as I am, if he need not come to me for work or wages."

That this is so, that at root the Eugenist is the Employer, there are multitudinous proofs on every side, but they are of necessity miscellaneous, and in many cases negative. The most enormous is in a sense the most negative: that no one seems able to imagine capitalist industrialism being sacrificed to any other object. By a curious recurrent slip in the mind, as irritating as a catch in a clock, people miss the main thing and concentrate on the mean thing. "Modern conditions" are treated as fixed, though the very word "modern" implies that they are fugitive. "Old ideas" are treated as impossible, though their very antiquity often proves their permanence. Some years ago some ladies petitioned that the platforms of our big railway stations should be raised, as it was more convenient for the hobble skirt. It never occurred to them to change to a sensible skirt. Still less did it occur to them that, compared with all the female fashions that have fluttered about on it, by this time St. Pancras is as historic as St. Peter's.

I could fill this book with examples of the universal, unconscious assumption that life and sex must live by the laws of "business" or industrialism, and not vice versa; examples from all the magazines, novels, and newspapers. In order to make it brief and typical, I take one case of a more or less Eugenist sort from a paper that lies open in front of me—a paper that still bears on its forehead the boast of being peculiarly an organ of democracy in revolt. To this a man writes to say that the spread of destitution will never be stopped until we have educated the lower classes in the methods by which the upper classes prevent procreation. The man had the horrible playfulness to sign his letter "Hopeful." Well, there are certainly many methods by which people in the upper classes prevent procreation; one of them is what used to be called "platonic friendship," till they found another name for it at the Old Bailey. I do not suppose the hopeful gentleman hopes for this; but some of us find the abortion he does hope for almost as abominable. That, however, is not the curious point. The curious point is that the hopeful one concludes by saying, "When people have large families and small wages, not only is there a high infantile death-rate, but often those who do live to grow up are stunted and weakened by having had to share the family income for a time with those who died early. There would be less unhappiness if there were no unwanted children." You will observe that he tacitly takes it for granted that the small wages and the income, desperately shared, are the fixed points, like day and night, the conditions of human life. Compared with them marriage and maternity are luxuries, things to be modified to suit the wage-market. There are unwanted children; but unwanted by whom? This man does not really mean that the parents do not want to have them. He means that the employers do not want to pay them properly. Doubtless, if you said to him directly, "Are you in favour of low wages?" he would say, "No." But I am not, in this chapter, talking about the effect on such modern minds of a cross-examination to which they do not subject themselves. I am talking about the way their minds work, the instinctive trick and turn of their thoughts, the things they assume before argument, and the way they faintly feel that the world is going. And, frankly, the turn of their mind is to tell the child he is not wanted, as the turn of my mind is to tell the profiteer he is not wanted. Motherhood, they feel, and a full childhood, and the beauty of brothers and sisters, are good things in their way, but not so good as a bad wage. About the mutilation of womanhood, and the massacre of men unborn, he signs himself "Hopeful." He is hopeful of female indignity, hopeful of human annihilation. But about improving the small bad wage he signs himself "Hopeless."

### Family to Fit Wages

Until the physically unfit are prevented [from] reproducing themselves, and the working classes restrict the numbers of their children (as the more educated classes do) to the capacity of the mother's physical strength to produce a few healthy offspring and the father's economic capacity to keep, educate, and put them out in the world, the miseries of overcrowded homes will continue.—LADY LAYLAND-BARRATT, BIRTH CONTROL NEWS, DEC, 1922

This is the first evidence of motive: the ubiquitous assumption that life and love must fit into a fixed framework of employment, even (as in this case) of bad employment. The second evidence is the tacit and total neglect of the scientific question in all the departments in which it is not an employment question; as, for instance, the marriages of the princely, patrician, or merely plutocratic houses. I do not mean, of course, that no scientific men have rigidly tackled these, though I do not recall any cases. But I am not talking of the merits of individual men of science, but of the push and power behind this movement, the thing that is able to make it fashionable and politically important. I say, if this power were an interest in truth, or even in humanity, the first field in which to study would be in the weddings of the wealthy. Not only would the records be more lucid, and the examples more in evidence, but the cases would be more interesting and more decisive. For the grand marriages have presented both extremes of the problem of pedigree—first the "breeding in and in," and later the most incongruous cosmopolitan blends. It would really be interesting to note which worked the best, or what point of compromise was safest. For the poor (about whom the newspaper Eugenists are always talking) cannot offer any test cases so complete. Waiters never had to marry waitresses, as princes had to marry princesses. And (for the other extreme) housemaids seldom marry Red Indians. It may be because there are none to marry. But to the millionaires the continents are flying railway stations, and the most remote races can be rapidly linked together. A marriage in London or Paris may chain Ravenna to Chicago, or Ben Cruachan to Bagdad. Many European aristocrats marry Americans, notoriously the most mixed stock in the world; so that the disinterested Eugenist, with a little trouble, might reveal rich stores of negro or Asiatic blood to his delighted employer. Instead of which he dulls our ears and distresses our refinement by tedious denunciations of the monochrome marriages of the poor.

### Negroes as a Savage Race

There is perhaps some connection between this obscure action and the disappearance of most savage races when brought into contact with high civilization, though there are other and well-known concomitant causes. But while most barbarous races disappear, some, like the negro, do not. —FRANCIS GALTON

For there is something really pathetic about the Eugenist's neglect of the aristocrat and his family affairs. People still talk about the pride of pedigree; but it strikes me as the one point on which the aristocrats are almost morbidly modest. We should be learned Eugenists if we were allowed to know half as much of their heredity as we are of their hairdressing. We see the modern aristocrat in the most human poses in the illustrated papers, playing, with his dog or parrot—nay, we see him playing with his child, or with his grandchild. But there is something heartrending in his refusal to play with his grandfather. There is often something vague and even fantastic about the antecedents of our most established families, which would afford the Eugenist admirable scope not only for investigation but for experiment. Certainly, if he could obtain the necessary powers, the Eugenist might bring off some startling effects with the mixed materials of the governing class. Suppose, to take wild and hypothetical examples, he were to marry a Scotch earl, say, to the daughter of a Jewish banker, or an English duke to an American parvenu of semi-Jewish extraction? What would happen? We have here an unexplored field.

It remains unexplored not merely through snobbery and cowardice, but because the Eugenist (at least the influential Eugenist) half-consciously knows it is no part of his job; what he is really wanted for is to get the grip of the governing classes on to the unmanageable output of poor people. It would not matter in the least if all Lord Cowdray's descendants grew up too weak to hold a tool or turn a wheel. It would matter very much, especially to Lord Cowdray, if all his employees grew up like that. The oligarch can be unemployable, because he will not be employed. Thus the practical and popular exponent of Eugenics has his face always turned towards the slums, and instinctively thinks in terms of them. If he talks of segregating some incurably vicious type of the sexual sort, he is thinking of a ruffian who assaults girls in lanes. He is not thinking of a millionaire like White, the victim of Thaw.[1] If he speaks of the hopelessness of feeble-mindedness, he is thinking of some stunted creature gaping at hopeless lessons in a poor school. He is not thinking of a millionaire like Thaw, the slayer of White. And this not because he is such a brute as to like people like White or Thaw any more than we do, but because he knows that his problem is the degeneration of the useful classes; because he knows that White would never have been a millionaire if all his workers had spent themselves on women as White did, that Thaw would never have been a millionaire if all his servants had been Thaws. The ornaments may be allowed to decay, but the machinery *must* be mended. That is the second proof of the plutocratic impulse behind all Eugenics: that no one thinks of applying it to the prominent classes. No one thinks of applying it where it could most easily be applied.

### Europe and Racial Hygiene

Europe is decaying, not only as a result of political cataclysms, but also because of a misconception of racial hygiene and a failure to counteract the forces of degeneration.—BIRTH CONTROL NEWS, DEC, 1922

A third proof is the strange new disposition to regard the poor as a *race;* as if they were a colony of Japs or Chinese coolies. It can be most clearly seen by comparing it with the old, more individual, charitable, and (as the Eugenists might say) sentimental view of poverty. In Goldsmith or Dickens or Hood there is a basic idea that the particular poor person ought not to be so poor: it is some accident or some wrong. Oliver Twist or Tiny Tim are fairy princes waiting for their fairy godmother. They are held as slaves, but rather as the hero and heroine of a Spanish or Italian romance were held as slaves by the Moors. The modern poor are getting to be regarded as slaves in the separate and sweeping sense of the negroes in the plantations. The bondage of the white hero to the black master was regarded as abnormal; the bondage of the black to the white master as normal. The Eugenist, for all I know, would regard the mere existence of

---

1. Editor: On June 25, 1906, Harry K. Thaw, a mentally deranged Pittsburgh millionaire, shot Stanford White, a wealthy New York City architect with a reputation for debauching teenage girls, including Thaw's wife, the former Evelyn Nesbit, when she was sixteen.

Tiny Tim as a sufficient reason for massacring the whole family of Cratchit; but, as a matter of fact, we have here a very good instance of how much more practically true to life is sentiment than cynicism. The poor are not a race or even a type. It is senseless to talk about breeding them; for they are not a breed. They are, in cold fact, what Dickens describes: "a dustbin of individual accidents," of damaged dignity, and often of damaged gentility. The class very largely consists of perfectly promising children, lost like Oliver Twist, or crippled like Tiny Tim. It contains very valuable things, like most dustbins. But the Eugenist delusion of the barbaric breed in the abyss affects even those more gracious philanthropists who almost certainly do want to assist the destitute and not merely to exploit them. It seems to affect not only their minds, but their very eyesight. Thus, for instance, Mrs. Alec Tweedie almost scornfully asks, "When we go through the slums, do we see beautiful children?" The answer is, "Yes, very often indeed." I have seen children in the slums quite pretty enough to be Little Nell or the outcast whom Hood called "young and so fair." Nor has the beauty anything necessarily to do with health; there are beautiful healthy children, beautiful dying children, ugly dying children, ugly uproarious children in Petticoat Lane or Park Lane. There are people of every physical and mental type, of every sort of health and breeding, in a single back street. They have nothing in common but the wrong we do them.

### Sex Without Children

Let the State give every defective full knowledge of how to avoid children, and supply them free with contraceptives; better still, offer them a substantial monetary reward to consent to sterilization, and as long as they are given to understand that the operation will "make no difference," they will in most cases agree.—*BIRTH CONTROL NEWS*, AUG, 1922

The important point is, however, that there is more fact and realism in the wildest and most elegant old fictions about disinherited dukes and long-lost daughters than there is in this Eugenist attempt to make the poor all of a piece—a sort of black fungoid growth that is ceaselessly increasing in a chasm. There is a cheap sneer at poor landladies: that they always say they have seen better days. Nine times out of ten they say it because it is true. What can be said of the great mass of Englishmen, by anyone who knows any history, except that they have seen better days? And the landlady's claim is not snobbish, but rather spirited; it is her testimony to the truth in the old tales of which I spoke: that she ought not to be so poor or so servile in status; that a normal person ought to have more property and more power in the State than *that*. Such dreams of lost dignity are perhaps the only things that stand between us and the cattle-breeding paradise now promised. Nor are such dreams by any means impotent. I remember Mr. T. P. O'Connor wrote an interesting article about Madame Humbert, in the course of which he said that Irish peasants, and probably most peasants, tended to have a half fictitious family legend about an estate to which they were entitled. This was written in the time when Irish peasants were landless in their land; and the delusion doubtless seemed all the more entertaining to the landlords who ruled them and the moneylenders who ruled the landlords. But the dream has conquered the realities. The phantom farms have materialised. Merely by tenaciously affirming the kind of pride that comes after a fall, by remembering the old civilisation and refusing the new, by recurring to an old claim that seemed to most Englishmen like the lie of a broken-down lodging-house keeper at Margate—by all this the Irish have got what they want, in solid mud and turf. That imaginary estate has conquered the Three Estates of the Realm.

But the homeless Englishman must not even remember a home. So far from his house being his castle, he must not have even a castle in the air. He must have no memories; that is why he is taught no history. Why is he told none of the truth about the mediaeval civilisation except a few cruelties and mistakes in chemistry? Why does a mediaeval burgher never appear till he can appear in a shirt and a halter? Why does a mediaeval monastery never appear till it is "corrupt" enough to shock the innocence of Henry VIII? Why do we hear of one charter—that of the barons—and not a word of the charters of the carpenters, smiths, shipwrights and all the rest? The reason is that the English peasant is not only not allowed to have an estate, he is not even allowed to have lost one. The past has to be painted pitch black, that it may be worse than the present.

### Race Cleansing

Obviously, it is this prodigious spawning of inferiors which must at all costs be prevented if society is to be saved from disruption and dissolution. Race cleansing is apparently the only thing that can stop it. Therefore, race cleansing must be our first concern.—LOTHROP STODDARD, *BIRTH CONTROL NEWS*, DEC, 1922

There is one strong, startling, outstanding thing about Eugenics, and that is its meanness, Wealth, and the social science supported by wealth, had tried an inhuman experiment. The experiment had entirely failed. They sought to make wealth accumulate—and they made men decay. Then, instead of confessing the error, and trying to restore the wealth, or attempting to repair the decay, they are trying to cover their first cruel experiment with a more cruel experiment. They put a poisonous plaster on a poisoned wound. Vilest of all, they actually quote the bewilderment produced among the poor by their first blunder as a reason for allowing them to blunder again. They are apparently ready to arrest all the opponents of their system as mad, merely because the system was maddening. Suppose a captain had collected volunteers in a hot, waste country by the assurance that he could lead them to water, and knew where to meet the rest of his regiment. Suppose he led them wrong, to a place where the regiment could not be for days, and there was no water. And suppose sunstroke struck them down on the sand man after man, and they kicked and danced and raved. And, when at last the regiment came, suppose the captain successfully concealed his mistake, because all his men had suffered too much from it to testify to its ever having occurred. What would you think of the gallant captain? It is pretty much what I think of this particular captain of industry.

Of course, nobody supposes that all Capitalists, or most Capitalists, are conscious of any such intellectual trick. Most of them are as much bewildered as the battered proletariat; but there are some who are less well-meaning and more mean. And these are leading their more generous colleagues towards the fulfilment of this ungenerous evasion, if not towards the comprehension of it. Now a ruler of the Capitalist civilisation, who has come to consider the idea of ultimately herding and breeding the workers like cattle, has certain contemporary problems to review. He has to consider what forces still exist in the modern world for the frustration of his design. The first question is how much remains of the old ideal of individual liberty. The second question is how far the modern mind is committed to such egalitarian ideas as may be implied in Socialism. The third is whether there is any power of resistance in the tradition of the populace itself. These three questions for the future I shall consider in their order in the final chapters that follow. It is enough to say here that I think the progress of these ideals has broken down at the precise point where they will fail to prevent the experiment. Briefly, the progress will have deprived the Capitalist of his old Individualist scruples, without committing him to his new Collectivist obligations. He is in a very perilous position; for he has ceased to be a Liberal without becoming a Socialist, and the bridge by which he was crossing has broken above an abyss of Anarchy.

### Marriage Without Children

If, then, society is ever to rid itself of its worse burdens, social reform must be increasingly supplemented by racial reform. Unfit individuals as well as unjust social conditions must be eliminated.

Even those persons who carry taints which make parenthood inadvisable need not be debarred from marriage. The sole limitation would be that they should have no children.—LOTHROP STODDARD, *BIRTH CONTROL NEWS*, DEC, 1922

# 14

# The Eclipse of Liberty

That is the problem, and that is why there is now no protection against
Eugenic or any other experiments. If the men who took away beer as an
unlawful pleasure had paused for a moment to define the lawful pleasures, there might be a different situation. If the men who had denied one
liberty had taken the opportunity to affirm other liberties, there might be
some defence for them. But it never occurs to them to admit any liberties at all. It never so much as crosses their minds. Hence the excuse for
the last oppression will always serve as well for the next oppression; and
to that tyranny there can be no end.

If such a thing as the Eugenic sociology had been suggested in the period
from Fox to Gladstone, it would have been far more fiercely repudiated
by the reformers than by the Conservatives. If Tories had regarded it as an
insult to marriage, Radicals would have far more resolutely regarded it as
an insult to citizenship. But in the interval we have suffered from a process resembling a sort of mystical parricide, such as is told of so many
gods, and is true of so many great ideas. Liberty has produced scepticism,
and scepticism has destroyed liberty. The lovers of liberty thought they
were leaving it unlimited, when they were only leaving it undefined. They
thought they were only leaving it undefined, when they were really leaving it undefended. Men merely finding themselves free found themselves
free to dispute the value of freedom. But the important point to seize
about this reactionary scepticism is that as it is bound to be unlimited in
theory, so it is bound to be unlimited in practice. In other words, the modern mind is set in an attitude which would enable it to advance, not only
towards Eugenic legislation, but towards any conceivable or inconceiva-
ble extravagances of Eugenics.

Those who reply to any plea for freedom invariably fall into a certain
trap. I have debated with numberless different people on these matters,

and I confess I find it amusing to see them tumbling into it one after another. I remember discussing it before a club of very active and intelligent Suffragists, and I cast it here for convenience in the form which it there assumed. Suppose, for the sake of argument, that I say that to take away a poor man's pot of beer is to take away a poor man's personal liberty, it is very vital to note what is the usual or almost universal reply. People hardly ever do reply, for some reason or other, by saying that a man's liberty consists of such and such things, but that beer is an exception that cannot be classed among them, for such and such reasons. What they almost invariably do say is something like this: "After all, what is liberty? Man must live as a member of a society, and must obey those laws which, etc., etc." In other words, they collapse into a complete confession that they *are* attacking all liberty and any liberty; that they *do* deny the very existence or the very possibility of liberty. In the very form of the answer they admit the full scope of the accusation against them. In trying to rebut the smaller accusation, they plead guilty to the larger one.

This distinction is very important, as can be seen from any practical parallel. Suppose we wake up in the middle of the night and find that a neighbour has entered the house not by the front-door but by the skylight; we may suspect that he has come after the fine old family jewellery. We may be reassured if he can refer it to a really exceptional event; as that he fell on to the roof out of an aeroplane, or climbed on to the roof to escape from a mad dog. Short of the incredible, the stranger the story the better the excuse; for an extraordinary event requires an extraordinary excuse. But we shall hardly be reassured if he merely gazes at us in a dreamy and wistful fashion and says, "After all, what is property? Why should material objects be thus artificially attached, etc., etc.?" We shall merely realise that his attitude allows of his taking the jewellery and everything else. Or if the neighbour approaches us carrying a large knife dripping with blood, we may be convinced by his story that he killed another neighbour in self-defence, that the quiet gentleman next door was really a homicidal maniac. We shall know that homicidal mania is exceptional and that we ourselves are so happy as not to suffer from it; and being free from the disease may be free from the danger. But it will not soothe us for the man with the gory knife to say softly and pensively, "After all, what is human life? Why should we cling to it? Brief at the best, sad at the brightest, it is itself but a disease from which, etc., etc.—" We shall perceive that the sceptic is in a mood not only to murder us but to massacre everybody in the street. Exactly the same effect which would be produced by the questions of "What is property?" and "What is life?" is produced by the question of "What is liberty?" It leaves the questioner free to disregard any liberty, or in other words to take any liberties. The very thing he says is an anticipatory excuse for anything he may choose to do. If he gags a man to prevent him from indulging in profane swearing, or locks him in the coal cellar to guard against his going on the spree, he can still be satisfied with saying, "After all, what is liberty? Man is a member of, etc., etc."

### Making a Marriage Null

The clause prohibiting marriage with a defective, which appeared in the 1912 Bill, was not reintroduced, nor was the new clause put down by Dr. Chapple to treat a marriage with a defective as null and void proceeded with.

—*EUGENICS REVIEW*, 1913

## No Protection Against Tyranny

That is the problem, and that is why there is now no protection against Eugenic or any other experiments. If the men who took away beer as an unlawful pleasure had paused for a moment to define the lawful pleasures, there might be a different situation. If the men who had denied one liberty had taken the opportunity to affirm other liberties, there might be some defence for them. But it never occurs to them to admit any liberties at all. It never so much as crosses their minds. Hence the excuse for the last oppression will always serve as well for the next oppression; and to that tyranny there can be no end.

Hence the tyranny has taken but a single stride to reach the secret and sacred places of personal freedom, where no sane man ever dreamed of seeing it; and especially the sanctuary of sex. It is as easy to take away a man's wife or baby as to take away his beer when you can say "What is liberty?"; just as it is as easy to cut off his head as to cut off his hair if you are free to say "What is life?" There is no rational philosophy of human rights generally disseminated among the populace, to which we can appeal in defence even of the most intimate or individual things that anybody can imagine. For so far as there was a vague principle in these things, that principle has been wholly changed. It used to be said that a man could have liberty, so long as it did not interfere with the liberty of others. This did afford some rough justification for the ordinary legal view of the man with the pot of beer. For instance, it was logical to allow some degree of distinction between beer and tea, on the ground that a man may be moved by excess of beer to throw the pot at somebody's head. And it may be said that the spinster is seldom moved by excess of tea to throw the tea-pot at anybody's head. But the whole ground of argument is now changed. For people do not consider what the drunkard does to others by throwing the pot, but what he does to himself by drinking the beer. The argument is based on health; and it is said that the Government must safeguard the health of the community. And the moment that is said, there ceases to be the shadow of a difference between beer and tea. People can certainly spoil their health with tea or with tobacco or with twenty other things. And there is no escape for the hygienic logician except to restrain and regulate them all. If he is to control the health of the community, he must necessarily control all the habits of all the citizens, and among the rest their habits in the matter of sex.

### Locking Up the Defective

A second Bill is now, however, the Mental Deficiency Act, 1913, which came into force on April 1st, 1914. Thus we may record, as the greatest achievement in the progress of modern eugenics, the coming into force in its decennial year of a beneficent measure which will, for the first time, take kindly care of the mentally defective as long as they need it, and in so doing will protect the future. The permanent care for which the Act provides is, under another name, the segregation which the principles of negative eugenics require in this case.—C.W. SALEEBY, 1914

## Beginning Where Despots Leave Off

But there is more than this. It is not only true that it is the last liberties of man that are being taken away; and not merely his first or most superficial liberties. It is also inevitable that the last liberties should be taken first. It is inevitable that the most private matters should be most under public coercion. This inverse variation is very important, though very little realised. If a man's personal health is a public concern, his most private acts are more public than his most public acts. The official must deal more directly with his cleaning his teeth in the morning than with his using his tongue in the market-place. The inspector must interfere *more* with how he sleeps in the middle of the night than with how he works in the course of the day. The private citizen must have much less to say about his bath or his bedroom window than about his vote or his banking account. The policeman must be in a new sense a private detective; and shadow him in private affairs rather than in public affairs. A policeman must shut doors behind him for fear he should sneeze, or shove pillows under him for fear he should snore. All this and things far more fantastic follow from the simple formula that the State must make itself responsible for the health of the citizen. But the point is that the policeman must deal primarily and promptly with the citizen in his relation to his home, and only indirectly and more doubtfully with the citizen in his relation to his city. By the whole logic of this test, the king must hear what is said in the inner chamber and hardly notice what is proclaimed from the housetops. We have heard of a revolution that turns everything upside down. But this is almost literally a revolution that turns everything inside out.

If a wary reactionary of the tradition of Metternich had wished in the nineteenth century to reverse the democratic tendency, he would naturally have begun by depriving the democracy of its margin of more dubious powers over more distant things. He might well begin, for instance, by removing the control of foreign affairs from popular assemblies; and there is a case for saying that a people may understand its own affairs, without knowing anything whatever about foreign affairs. Then he might centralise great national questions, leaving a great deal of local government in local questions. This would proceed so for a long time before it occurred to the blackest terrorist of the despotic ages to interfere with a man's own habits in his own house. But the new sociologists and legislators are, by the nature of their theory, bound to begin where the despots leave off, even if they leave off where the despots begin. For them, as they would put it, the first things must be the very fountains of life, love and birth and babyhood; and these are always covered fountains, flowing in the quiet courts of the home. For them, as Mr. H. G. Wells put it, life itself may be regarded merely as a tissue of births. Thus they are coerced by their own rational principle to begin all coercion at the other end; at the inside end. What happens to the outside end, the external and remote powers of the citizen, they do not very much care; and it is probable that the democratic institutions of recent centuries will be allowed to decay in undisturbed dignity for a century or two more. Thus our civilisation will find itself in an interesting situation, not without humour; in which the citizen is still supposed to wield imperial powers over the ends of the earth, but has admittedly no power over his own body and soul at all. He will still be consulted by politicians about whether opium is good for Chinamen, but not about whether ale is good for him. He will be cross-examined for his opinions about the danger of allowing Kamskatka to have a war-fleet, but not about allowing his own child to have a wooden sword. About all, he will be consulted about the delicate diplomatic crisis created by the proposed marriage of the Emperor of China, and not allowed to marry as he pleases.

Part of this prophecy or probability has already been accomplished; the rest of it, in the absence of any protest, is in process of accomplishment. It would be easy to give an almost endless catalogue of examples, to show how, in dealing with the poorer classes at least, coercion has already come near to a direct control of the relations of the sexes. But I am much more concerned in this chapter to point out that all these things have been adopted in principle, even where they have not been adopted in practice. It is much more vital to realise that the reformers have possessed themselves of a principle, which will cover all such things if it be granted, and which is not sufficiently comprehended to be contradicted. It is a principle whereby the deepest things of flesh and spirit must have the most direct relation with the dictatorship of the State. They must have it, by the whole reason and rationale upon which the thing depends. It is a system that might be symbolised by the telephone from headquarters standing by a man's bed. He must have a relation to Government like his relation to God. That is, the more he goes into the inner chambers, and the more he closes the doors, the more he is alone with the law. The social machinery which makes such a State uniform and submissive will be worked outwards from the household as from a handle, or a single mechanical knob or button. In a horrible sense, loaded with fear and shame and every detail of dishonour, it will be true to say that charity begins at home.

Charity will begin at home in the sense that all home children will be like charity children. Philanthropy will begin at home, for all household-ers will be like paupers. Police administration will begin at home, for all citizens will be like convicts. And when health and the humours of daily life have passed into the domain of this social discipline, when it is admitted that the community must primarily control the primary habits, when all law begins, so to speak, next to the skin or nearest the vitals—then indeed it will appear absurd that marriage and maternity should not be similarly ordered. Then indeed it will seem to be illogical, and it will be illogical, that love should be free when life has lost its freedom.

### Using Drugs to Control

The dependence of emotional disposition upon the ductless glands, said Mr. Russell, was a discovery of great importance, which would in time make it possible to produce artificially any disposition desired by Governments.
—BERTRAND RUSSELL, *BIRTH CONTROL NEWS*, FEB. 1924

## English Liberties Destroyed

So passed, to all appearance, from the minds of men the strange dream and fantasy called freedom. Whatever be the future of these evolutionary experiments and their effect on civilisation, there is one land at least that has something to mourn. For us in England something will have perished which our fathers valued all the more because they hardly troubled to name it; and whatever be the stars of a more universal destiny, the great star of our night has set. The English had missed many other things that men of the same origins had achieved or retained. Not to them was given, like the French, to establish eternal communes and clear codes of equality; not to them, like the South Germans, to keep the popular culture of their songs; not to them, like the Irish, was it given to die daily for a great religion. But a spirit had been with them from the first which fenced, with a hundred quaint customs and legal fictions, the way of a man who wished to walk nameless and alone. It was not for nothing that they forgot all their laws to remember the name of an outlaw, and filled the green heart of England with the figure of Robin Hood. It was not for nothing that even their princes of art and letters had about them something of kings incognito, undiscovered by formal or academic fame; so that no eye can follow the young Shakespeare as he came up the green lanes from Stratford, or the young Dickens when he first lost himself among the lights of London. It is not for nothing that the very roads are crooked and capricious, so that a man looking down on a map like a snaky labyrinth, could tell that he was looking on the home of a wandering people. A spirit at once wild and familiar rested upon its woodlands like a wind at rest. If that spirit be indeed departed, it matters little that it has been driven out by perversions it had itself permitted, by monsters it had idly let loose. Industrialism and Capitalism and the rage for physical science were English experiments in the sense that the English lent themselves to their encouragement; but there was something else behind them and within them that was not they—its name was liberty, and it was our life. It may be that this delicate and tenacious spirit has at last evaporated. If so, it matters little what becomes of the external experiments of our nation in later time. That at which we look will be a dead thing alive with its own parasites. The English will have destroyed England.

### Sterilize Criminals

Mr. Justice Roche remarked that some time it might be a part of the English law to sterilize people with such tendencies as the prisoner, and the sooner English doctors studied the question the better and the sooner we were likely to have a different type of people to deal with.—*BIRTH CONTROL NEWS*, NOV. 1922

# 15

# The Transformation of Socialism

In short, people decided that it was impossible to achieve any of the good of Socialism, but they comforted themselves by achieving all the bad. All that official discipline, about which the Socialists themselves were in doubt or at least on the defensive, was taken over bodily by the Capitalists. They have now added all the bureaucratic tyrannies of a Socialist state to the old plutocratic tyrannies of a Capitalist State. For the vital point is that it did not in the smallest degree diminish the inequalities of a Capitalist State. It simply destroyed such individual liberties as remained among its victims. It did not enable any man to build a better house; it only limited the houses he might live in—or how he might manage to live there; forbidding him to keep pigs or poultry or to sell beer or cider.

Socialism is one of the simplest ideas in the world. It has always puzzled me how there came to be so much bewilderment and misunderstanding and miserable mutual slander about it. At one time I agreed with Socialism, because it was simple. Now I disagree with Socialism, because it is too simple. Yet most of its opponents still seem to treat it, not merely as an iniquity but as a mystery of iniquity, which seems to mystify them even more than it maddens them. It may not seem strange that its antagonists should be puzzled about what it is. It may appear more curious and interesting that its admirers are equally puzzled. Its foes used to denounce Socialism as Anarchy, which is its opposite. Its friends seemed to suppose that it is a sort of optimism, which is almost as much of an opposite. Friends and foes alike talked as if it involved a sort of faith in ideal human nature; why I could never imagine. The Socialist system, in a more special sense than any other, is founded not on optimism but on original sin. It proposes that the State, as the conscience of the community, should possess all primary forms of property; and that obviously on the ground that men cannot be trusted to own or barter or combine or compete without injury to themselves. Just as a State might own all the guns lest people should shoot each other, so this State would own all the gold and land lest they should cheat or rackrent or exploit each other. It seems extraordinarily simple and even obvious; and so it is. It is too obvious to be true. But while it is obvious, it seems almost incredible that anybody ever thought it optimistic.

I am myself primarily opposed to Socialism, or Collectivism or Bolshevism or whatever we call it, for a primary reason not immediately involved here: the ideal of Property. I say the ideal and not merely the idea; and this alone disposes of the moral mistake in the matter. It disposes of all the dreary doubts of the Anti-Socialists about men not yet being angels, and all the yet drearier hopes of the Socialists about men soon being supermen. I do not admit that private property is a concession to baseness and selfishness; I think it is a point of honour. I think it is the most truly popular of all points of honour. But this, though it has everything to do with my plea for a domestic dignity, has nothing to do with this passing summary of the situation of Socialism. I only remark in passing that it is vain for the more vulgar sort of Capitalist, sneering at ideals, to say to me that in order to have Socialism "You must alter human nature." I answer "Yes. You must alter it for the worse."

The clouds were considerably cleared away from the meaning of Socialism by the Fabians of the 'nineties; by Mr. Bernard Shaw, a sort of anti-romantic Quixote, who charged chivalry as chivalry charged windmills, with Sidney Webb for his Sancho Panza. In so far as these paladins had a castle to defend, We may say that their castle was the Post Office. The red pillar-box was the immovable post against which the irresistible force of Capitalist individualism was arrested. Business men who said that nothing could be managed by the State were forced to admit that they trusted all their business letters and business telegrams to the State.

After all, it was not found necessary to have an office competing with another office, trying to send out pinker postage stamps or more picturesque postmen. It was not necessary to efficiency that the postmistress should buy a penny stamp for a halfpenny and sell it for twopence; or that she should haggle and beat customers down about the price of a postal order; or that she should always take tenders for telegrams. There was obviously nothing actually impossible about the State management of national needs; and the Post Office was at least tolerably managed. Though it was not always a model employer, by any means, it might be made so by similar methods. It was not impossible that equitable pay, and even equal pay, could be given to the Postmaster-General and the postman. We had only to extend this rule of public responsibility, and we should escape from all the terror of insecurity and torture of compassion, which hag-rides humanity in the insane extremes of economic inequality and injustice. As Mr. Shaw put it, "A man must save Society's honour before he can save his own."

That was one side of the argument: that the change would remove inequality; and there was an answer on the other side. It can be stated most truly by putting another model institution and edifice side by side with the Post Office. It is even more of an ideal republic, or commonwealth without competition or private profit. It supplies its citizens not only with the stamps but with clothes and food and lodging, and all they require. It observes considerable level of equality in these things; notably in the clothes. It not only supervises the letters but all the other human communications; notably the sort of evil communications that corrupt good manners. This twin model to the Post Office is called the Prison. And much of the scheme for a model State was regarded by its opponents as a scheme for a model prison; good because it fed men equally, but less acceptable since it imprisoned them equally.

It is better to be in a bad prison than in a good one. From the standpoint of the prisoner this is not at all a paradox; if only because in a bad prison he is more likely to escape. But apart from that, a man was in many ways better off in the old dirty and corrupt prison, where he could bribe turnkeys to bring him drink and meet fellow-prisoners to drink with. Now that is exactly the difference between the present system and the proposed system. Nobody worth talking about respects the present system. Capitalism is a corrupt prison. That is the best that can be said for Capitalism. But it is something to be said for it; for a man is a little freer in that corrupt prison than he would be in a complete prison. As a man can find one jailer more lax than another, so he could find one employer more kind than another; he has at least a choice of tyrants. In the other case he finds the same tyrant at every turn. Mr. Shaw and other rational Socialists have agreed that the State would be in practice government by a small group. Any independent man who disliked that group would find his foe waiting for him at the end of every road.

It may be said of Socialism, therefore, very briefly, that its friends recommended it as increasing equality, while its foes resisted it as decreasing liberty. On the one hand it was said that the State could provide homes and meals for all; on the other it was answered that this could only be done by State officials who would inspect houses and regulate meals. The compromise eventually made was one of the most interesting and even curious cases in history. It was decided to do everything that had ever been denounced in Socialism, and nothing that had ever been desired in it. Since it was supposed to gain equality at the sacrifice of liberty, we proceeded to prove that it was possible to sacrifice liberty without gaining equality. Indeed, there was not the faintest attempt to gain equality, least of all economic equality. But there was a very spirited and vigorous effort to eliminate liberty, by means of an entirely new crop of crude regulations and interferences. But it was not the Socialist State regulating those whom it fed, like children or even like convicts. It was the Capitalist State raiding those whom it had trampled and deserted in every sort of den, like outlaws or broken men. It occurred to the wiser sociologists that, after all, it would be easy to proceed more promptly to the main business of bully-ing men, without having gone through the laborious preliminary business of supporting them. After all, it was easy to inspect the house without having helped to build it; it was even possible, with luck, to inspect the house in time to prevent it being built. All that is described in the docu-ments of the Housing Problem; for the people of this age loved problems and hated solutions. It was easy to restrict the diet without providing the dinner. All that can be found in the documents of what is called Temper-ance Reform.

In short, people decided that it was impossible to achieve any of the good of Socialism, but they comforted themselves by achieving all the bad. All that official discipline, about which the Socialists themselves were in doubt or at least on the defensive, was taken over bodily by the Capitalists. They have now added all the bureaucratic tyrannies of a Socialist state to the old plutocratic tyrannies of a Capitalist State. For the vital point is that it did not in the smallest degree diminish the inequalities of a Capitalist State. It simply destroyed such individual liberties as remained among its victims. It did not enable any man to build a better house; it only limited the houses he might live in—or how he might man-age to live there; forbidding him to keep pigs or poultry or to sell beer or cider. It did not even add anything to a man's wages; it only took away something from a man's wages and locked it up, whether he liked it or not, in a sort of money-box which was regarded as a medicine-chest. It does not send food into the house to feed the children; it only sends an inspector into the house to punish the parents for having no food to feed them. It does not see that they have got a fire; it only punishes them for not having a fireguard. It does not even occur to it to provide the fire-guard.

Now this anomalous situation will probably ultimately evolve into the Servile State of Mr. Belloc's thesis. The poor will sink into slavery; it might as correctly be said that the poor will rise into slavery. That is to say, sooner or later, it is very probable that the rich will take over the philanthropic as well as the tyrannic side of the bargain; and will feed men like slaves as well as hunting them like outlaws. But for the purpose of my own argument it is not necessary to carry the process so far as this, or indeed any farther than it has already gone. The purely negative stage of interference, at which we have stuck for the present, is in itself quite favourable to all these eugenical experiments. The capitalist whose half-conscious thought and course of action I have simplified into a story in the preceding chapters, finds this insufficient solution quite sufficient for his purposes. What he has felt for a long time is that he must check or improve the reckless and random breeding of the submerged race, which is at once outstripping his requirements and failing to fulfil his needs. Now the anomalous situation has already accustomed him to stopping things. The first interferences with sex need only be negative; and there are already negative interferences without number. So that the study of this stage of Socialism brings us to the same conclusion as that of the ideal of liberty as formally professed by Liberalism. The ideal of liberty is lost, and the ideal of Socialism is changed, till it is a mere excuse for the oppression of the poor.

The first movements for intervention in the deepest domestic concerns of the poor all had this note of negative interference. Official papers were sent round to the mothers in poor streets; papers in which a total stranger asked these respectable women questions which a man would be killed for asking in the class of what were called gentlemen or in the countries of what were called free men. They were questions supposed to refer to the conditions of maternity; but the point is here that the reformers did not begin by building up those economic or material conditions. They did not attempt to pay money or establish property to create those conditions. They never give anything—except orders. Another form of the intervention, and one already mentioned, is the kidnapping of children upon the most fantastic excuses of sham psychology. Some people established an apparatus of tests and trick questions; which might make an amusing game of riddles for the family fireside, but seems an insufficient reason for mutilating and dismembering the family. Others became interested in the hopeless moral condition of children born in the economic condition which they did not attempt to improve. They were great on the fact that crime was a disease; and carried on their criminological studies so successfully as to open the reformatory for little boys who played truant; there was no reformatory for reformers. I need not pause to explain that crime is not a disease. It is criminology that is a disease.

Finally one thing may be added which is at least clear. Whether or no the organisation of industry will issue positively in a eugenical reconstruction of the family, it has already issued negatively, as in the negations already noted, in a partial destruction of it. It took the form of a propaganda of popular divorce, calculated at least to accustom the masses to a new notion of the shifting and re-grouping of families. I do not discuss the question of divorce here, as I have done elsewhere, in its intrinsic character; I merely note it as one of these negative reforms which have been substituted for positive economic equality. It was preached with a weird hilarity, as if the suicide of love were something not only humane but happy. But it need not be explained, and certainly it need not be denied, that the harassed poor of a diseased industrialism were indeed maintaining marriage under every disadvantage, and often found individual relief in divorce. Industrialism does produce many unhappy marriages, for the same reason that it produces so many unhappy men. But all the reforms were directed to rescuing the industrialism rather than the happiness. Poor couples were to be divorced because they were already divided. Through all this modern muddle there runs the curious principle of sacrificing the ancient uses of things because they do not fit in with the modern abuses. When the tares are found in the wheat, the greatest promptitude and practicality is always shown in burning the wheat and gathering the tares into the barn. And since the serpent coiled about the chalice had dropped his poison in the wine of Cana, analysts were instantly active in the effort to preserve the poison and to pour away the wine.

### Sterilization Before Marriage in Oregon

Under its terms, all applicants for marriage licenses must prove that they have at least the mentality of a child of twelve years and be free from communicable or contagious diseases. Failure to pass the examination would preclude the issuance of a license to marry unless one or both of the parties submitted to sterilization.—BIRTH CONTROL NEWS, APR. 1923

### In California Only the Fit to Marry

Birth control, limitation of marriage to the fit and sterilization of the criminal and insane were unqualifiedly endorsed by Dr. H. G. Brainard of Los Angeles, president of the California State Medical Association, in an address today at the opening session of the Association's annual convention.
—BIRTH CONTROL NEWS, AUG. 1923

CHAPTER

# 16

## The End of Household Goods

The working classes have no reserves of property with which to defend their relics of religion. They have no religion with which to sanctify and dignify their property. Above all, they are under the enormous disadvantage of being right without knowing it. They hold their sound principles as if they were sullen prejudices. They almost secrete their small property as if it were stolen property. Often a poor woman will tell a magistrate that she sticks to her husband, with the defiant and desperate air of a wanton resolved to run away from her husband. Often she will cry as hopelessly, and as it were helplessly, when deprived of her child as if she were a child deprived of her doll.

The only place where it is possible to find an echo of the mind of the English masses is either in conversation or in comic songs. The latter are obviously the more dubious; but they are the only things recorded and quotable that come anywhere near it. We talk about the popular Press; but in truth there is no popular Press. It may be a good thing; but, anyhow, most readers would be mildly surprised if a newspaper leading article were written in the language of a navvy. Sometimes the Press is interested in things in which the democracy is also genuinely interested; such as horse-racing. Sometimes the Press is about as popular as the Press Gang. We talk of Labour leaders in Parliament; but they would be highly unparliamentary if they talked like labourers. The Bolshevists, I believe, profess to promote something that they call "proletarian art," which only shows that the word Bolshevism can sometimes be abbreviated into bosh. That sort of Bolshevist is not a proletarian, but rather the very thing he accuses everybody else of being. The Bolshevist is above all a bourgeois; a Jewish intellectual of the town. And the real case against industrial intellectualism could hardly be put better than in this very comparison.

There has never been such a thing as proletarian art; but there has emphatically been such a thing as peasant art. And the only literature which even reminds us of the real tone and talk of the English working classes is to be found in the comic song of the English music-hall.

I first heard one of them on my voyage to America, in the midst of the sea within sight of the New World, with the Statue of Liberty beginning to loom up on the horizon. From the lips of a young Scotch engineer, of all people in the world, I heard for the first time these immortal words from a London music-hall song:—

"Father's got the sack from the water-works

For smoking of his old cherry-briar;

Father's got the sack from the water-works

'Cos he might set the water-works on fire."

As I told my friends in America, I think it no part of a patriot to boast; and boasting itself is certainly not a thing to boast of. I doubt the persuasive power of English as exemplified in Kipling, and one can easily force it on foreigners too much, even as exemplified in Dickens. I am no Imperialist, and only on rare and proper occasions a jingo. But when I hear those words about Father and the water-works, when I hear under far-off foreign skies anything so gloriously English as that, then indeed (I said to them), then indeed:

"I thank the goodness and the grace

That on my birth have smiled,

And made me, as you see me here

A little English child."

But that noble stanza about the water-works has other elements of nobility besides nationality. It provides a compact and almost perfect summary of the whole social problem in industrial countries like England and America. If I wished to set forth systematically the elements of the ethical and economic problem in Pittsburgh or Sheffield, I could not do better than take these few words as a text, and divide them up like the heads of a sermon. Let me note the points in some rough fashion here.

1.—**Father**. This word is still in use among the more ignorant and ill-paid of the industrial community; and is the badge of an old convention or unit called the family. A man and woman having vowed to be faithful to each other, the man makes himself responsible for all the children of the woman, and is thus generically called "Father." It must not be supposed that the poet or singer is necessarily one of the children. It may be the wife, called by the same ritual "Mother." Poor English wives say "Father" as poor Irish wives say "Himself," meaning the titular head of the house. The point to seize is that among the ignorant this convention or custom still exists. Father and the family are the foundations of thought; the natural authority still comes natural to the poet; but it is overlaid and thwarted with more artificial authorities; the official, the schoolmaster, the policeman, the employer, and so on. What these forces fighting the family are we shall see, my dear brethren, when we pass to our second heading; which is:

**2.—Got the Sack**. This idiom marks a later stage of the history of the language than the comparatively primitive word "Father." It is needless to discuss whether the term comes from Turkey or some other servile society. In America they say that Father has been fired. But it involves the whole of the unique economic system under which Father has now to live. Though assumed by family tradition to be a master, he can now, by industrial tradition, only be a particular kind of servant, a servant who has not the security of a slave. If he owned his own shop and tools, he could not get the sack. If his master owned him, he could not get the sack. The slave and the guildsman know where they will sleep every night; it was only the proletarian of individualist industrialism who could get the sack, if not in the style of the Bosphorus, at least in the sense of the Embankment. We pass to the third heading.

**3.—From the Water-works.** This detail of Father's life is very important; for this is the reply to most of the Socialists, as the last section is to so many of the Capitalists. The water-works which employed Father is a very large, official and impersonal institution. Whether it is technically a bureaucratic department or a big business makes little or no change in the feelings of Father in connection with it. The water-works might or might not be nationalised; and it would make no necessary difference to Father being fired, and no difference at all to his being accused of playing with fire. In fact, if the Capitalists are more likely to give him the sack, the Socialists are even more likely to forbid him the smoke. There is no freedom for Father except in some sort of private ownership of things like water and fire. If he owned his own well his water could never be cut off, and while he sits by his own fire his pipe can never be put out. That is the real meaning of property, and the real argument against Socialism; probably the only argument against Socialism.

**4.—For Smoking.** Nothing marks this queer intermediate phase of industrialism more strangely than the fact that, while employers still claim the right to sack him like a stranger, they are already beginning to claim the right to supervise him like a son. Economically he can go and

starve on the Embankment; but ethically and hygienically he must be controlled and coddled in the nursery. Government repudiates all responsibility for seeing that he gets bread. But it anxiously accepts all responsibility
for seeing that he does not get beer. It passes an Insurance Act to force
him to provide himself with medicine; but it is avowedly indifferent to
whether he is able to provide himself with meals. Thus while the sack is
inconsistent with the family, the supervision is really inconsistent with the
sack. The whole thing is a tangled chain of contradictions. It is true that in
the special and sacred text of scripture we are here considering, the smoking is forbidden on a general and public and not on a medicinal and private ground. But it is none the less relevant to remember that, as his
masters have already proved that alcohol is a poison, they may soon prove
that nicotine is a poison. And it is most significant of all that this sort of
danger is even greater in what is called the new democracy of America
than in what is called the old oligarchy of England. When I was in America, people were already "defending" tobacco. People who defend
tobacco are on the road to proving that daylight is defensible, or that it is
not really sinful to sneeze. In other words, they are quietly going mad.

**5.—Of his old Cherry-briar.** Here we have the intermediate and
anomalous position of the institution of Property. The sentiment still
exists, even among the poor, or perhaps especially among the poor. But it
is attached to toys rather than tools; to the minor products rather than to
the means of production. But something of the sanity of ownership is still
to be observed; for instance, the element of custom and continuity. It was
an *old* cherry-briar; systematically smoked by Father in spite of all wiles
and temptations to Woodbines and gaspers; an old companion possibly
connected with various romantic or diverting events in Father's life. It is
perhaps a relic as well as a trinket. But because it is not a true tool,
because it gives the man no grip on the creative energies of society, it is,
with all the rest of his self-respect, at the mercy of the thing called the
sack. When he gets the sack from the water-works, it is only too probable
that he will have to pawn his old cherry-briar.

**6.—'Cos he might set the water-works on fire.** And that single line,
like the lovely single lines of the great poets, is so full, so final, so perfect
a picture of all the laws we pass and all the reasons we give for them, so
exact an analysis of the logic of all our precautions at the present time,
that the pen falls even from the hands of the commentator; and the mas-
terpiece is left to speak for itself.

Some such analysis as the above gives a better account than most of
the anomalous attitude and situation of the English proletarian to-day. It is
the more appropriate because it is expressed in the words he actually uses;

which certainly do not include the word "proletarian." It will be noted that everything that goes to make up that complexity is in an unfinished state. Property has not quite vanished; slavery has not quite arrived; marriage exists under difficulties; social regimentation exists under restraints, or rather under subterfuges. The question which remains is which force is gaining on the other, and whether the old forces are capable of resisting the new. I hope they are; but I recognise that they resist under more than one heavy handicap. The chief of these is that the family feeling of the workmen is by this time rather an instinct than an ideal. The obvious thing to protect an ideal is a religion. The obvious thing to protect the ideal of marriage is the Christian religion. And for various reasons, which only a history of England could explain (though it hardly ever does), the working classes of this country have been very much cut off from Christianity. I do not dream of denying, indeed I should take every opportunity of affirming, that monogamy and its domestic responsibilities can be defended on rational apart from religious grounds. But a religion is the practical protection of any moral idea which has to be popular and which has to be pugnacious. And our ideal, if it is to survive, will have to be both.

Those who make merry over the landlady who has seen better days, of whom something has been said already, commonly speak, in the same jovial journalese, about her household goods as her household gods. They would be much startled if they discovered how right they are. Exactly what is lacking to the modern materialist is something that can be what the household gods were to the ancient heathen. The household gods of the heathen were not only wood and stone; at least there is always more than that in the stone of the hearth-stone and the wood of the roof-tree. So long as Christianity continued the tradition of patron saints and portable relics, this idea of a blessing on the household could continue. If men had not domestic divinities, at least they had divine domesticities. When Christianity was chilled with Puritanism and rationalism, this inner warmth or secret fire in the house faded on the hearth. But some of the embers still glow or at least glimmer; and there is still a memory among the poor that their material possessions are something sacred. I know poor men for whom it is the romance of their lives to refuse big sums of money for an old copper warming-pan. They do not want it, in any sense of base utility. They do not use it as a warming-pan; but it warms them for all that. It is indeed, as Sergeant Buzfuz humorously observed, a cover for hidden fire. And the fire is that which burned before the strange and uncouth wooden gods, like giant dolls, in the huts of ancient Italy. It is a household god. And I can imagine some such neglected and unlucky English man dying with his eyes on the red gleam of that piece of copper, as happier men have died with their eyes on the golden gleam of a chalice or a cross.

It will thus be noted that there has always been some connection between a mystical belief and the materials of domesticity; that they generally go together; and that now, in a more mournful sense, they are gone together. The working classes have no reserves of property with which to defend their relics of religion. They have no religion with which to sanctify and dignify their property. Above all, they are under the enormous disadvantage of being right without knowing it. They hold their sound principles as if they were sullen prejudices. They almost secrete their small property as if it were stolen property. Often a poor woman will tell a magistrate that she sticks to her husband, with the defiant and desperate air of a wanton resolved to run away from her husband. Often she will cry as hopelessly, and as it were helplessly, when deprived of her child as if she were a child deprived of her doll. Indeed, a child in the street, crying for her lost doll, would probably receive more sympathy than she does.

Meanwhile the fun goes on; and many such conflicts are recorded, even in the newspapers, between heart-broken parents and house-breaking philanthropists; always with one issue, of course. There are any number of them that never get into the newspapers. And we have to be flippant about these things as the only alternative to being rather fierce; and I have no desire to end on a note of universal ferocity. I know that many who set such machinery in motion do so from motives of sincere but confused compassion, and many more from a dull but not dishonourable medical or legal habit. But if I and those who agree with me tend to some harshness and abruptness of condemnation, these worthy people need not be altogether impatient with our impatience. It is surely beneath them, in the scope of their great schemes, to complain of protests so ineffectual about wrongs so individual. I have considered in this chapter the chances of general democratic defence of domestic honour, and have been compelled to the conclusion that they are not at present hopeful; and it is at least clear that we cannot be founding on them any personal hopes. If this conclusion leaves us defeated we submit that it leaves us disinterested. Ours is not the sort of protest, at least, that promises anything even to the demagogue, let alone the sycophant. Those we serve will never rule, and those we pity will never rise. Parliament will never be surrounded by a mob of submerged grandmothers brandishing pawn-tickets. There is no trade union of defective children. It is not very probable that modern government will be overturned by a few poor dingy devils who are sent to prison by mistake, or rather by ordinary accident. Surely it is not for those magnificent Socialists, or those great reformers and recon-structors of Capitalism, sweeping onward to their scientific triumphs and caring for none of these things, to murmur at our vain indignation. At least if it is vain it is the less venal; and in so far as it is hopeless it is also thankless. They have their great campaigns and cosmopolitan systems for the regimentation of millions, and the records of science and progress. They need not be angry with us, who plead for those who will never read our words or reward our effort, even with gratitude. They need surely have no worse mood towards us than mystification, seeing that in recall-ing these small things of broken hearts or homes, we are but recording what cannot be recorded; trivial tragedies that will fade faster and faster in the flux of time, cries that fail in a furious and infinite wind, wild words of despair that are written only upon running water; unless, indeed, as some so stubbornly and strangely say, they are somewhere cut deep into a rock, in the red granite of the wrath of God.

### A Soulless Humanity

Most persons seem to have a vague idea that a new element, especially fash-ioned in heaven, and not transmitted by simple descent, is introduced into the body of every newly born infant. Such a notion is unfitted to stand upon any scientific basis with which we are acquainted.—FRANCIS GALTON, 1865

# 17

# A Short Chapter

With whom, alas, did England go to war? England went to war with the
Superman in his native home. She went to war with that very land of scientific culture from which the very ideal of a Superman had come.

Round about the year 1913 Eugenics was turned from a fad to a fashion.
Then, if I may so summarise the situation, the joke began in earnest. The
organising mind which we have seen considering the problem of slum
population, the popular material and the possibility of protests, felt that
the time had come to open the campaign. Eugenics began to appear in big
headlines in the daily Press, and big pictures in the illustrated papers. A
foreign gentleman named Bolce, living at Hampstead, was advertised on
a huge scale as having every intention of being the father of the Superman. It turned out to be a Superwoman, and was called Eugenette. The
parents were described as devoting themselves to the production of perfect prenatal conditions. They "eliminated everything from their lives
which did not tend towards complete happiness." Many might indeed be
ready to do this; but in the voluminous contemporary journalism on the
subject I can find no detailed notes about how it is done. Communications
were opened with Mr. H. G. Wells, with Dr. Saleeby, and apparently with
Dr. Karl Pearson. Every quality desired in the ideal baby was carefully
cultivated in the parents. The problem of a sense of humour was felt to be
a matter of great gravity. The Eugenist couple, naturally fearing they
might be deficient on this side, were so truly scientific as to have resort to
specialists. To cultivate a sense of fun, they visited Harry Lauder, and
then Wilkie Bard, and afterwards George Robey; but all, it would appear,
in vain. To the newspaper reader, however, it looked as if the names of
Metchnikoff and Steinmetz and Karl Pearson would soon be quite as
familiar as those of Robey and Lauder and Bard. Arguments about these
Eugenic authorities, reports of the controversies at the Eugenic Congress,

filled countless columns. The fact that Mr. Bolce, the creator of perfect pre-natal conditions, was afterwards sued in a law-court for keeping his own flat in conditions of filth and neglect, cast but a slight and momentary shadow upon the splendid dawn of the science. It would be vain to record any of the thousand testimonies to its triumph. In the nature of things, this should be the longest chapter in the book, or rather the beginning of another book. It should record, in numberless examples, the triumphant popularisation of Eugenics in England. But as a matter of fact this is not the first chapter but the last. And this must be a very short chapter, because the whole of this story was cut short. A very curious thing happened. England went to war.

This would in itself have been a sufficiently irritating interruption in the early life of Eugenette, and in the early establishment of Eugenics. But a far more dreadful and disconcerting fact must be noted. With whom, alas, did England go to war? England went to war with the Superman in his native home. She went to war with that very land of scientific culture from which the very ideal of a Superman had come. She went to war with the whole of Dr. Steinmetz, and presumably with at least half of Dr. Karl Pearson. She gave battle to the birthplace of nine-tenths of the professors who were the prophets of the new hope of humanity. In a few weeks the very name of a professor was a matter for hissing and low plebeian mirth. The very name of Nietzsche, who had held up this hope of something superhuman to humanity, was laughed at for all the world as if he had been touched with lunacy. A new mood came upon the whole people; a mood of marching, of spontaneous soldierly vigilance and democratic discipline, moving to the faint tune of bugles far away. Men began to talk strangely of old and common things, of the counties of England, of its quiet landscapes, of motherhood and the half-buried religion of the race. Death shone on the land like a new daylight, making all things vivid and visibly dear. And in the presence of this awful actuality it seemed, somehow or other, as if even Mr. Bolce and the Eugenic baby were things unaccountably far-away and almost, if one may say so, funny.

### Strong Hampered by Unfit

The aim of Nietzsche was to evolve great men in all the walks of life and to remove all restraint which would impede their evolution. And in modern life he saw that the strong and healthy were hampered by the care of the unfit and the weak. So Nietzsche blamed the Christian religion.—*Eugenics Review, 1913*

Such a revulsion requires explanation, and it may be briefly given. There was a province of Europe which had carried nearer to perfection than any other the type of order and foresight that are the subject of this book. It had long been the model State of all those more rational moralists who saw in science the ordered salvation of society. It was admittedly ahead of all other States in social reform. All the systematic social reforms were professedly and proudly borrowed from it. Therefore when this province of Prussia found it convenient to extend its imperial system to the neighbouring and neutral State of Belgium, all these scientific enthusiasts had a privilege riot always granted to mere theorists. They had the gratification of seeing their great Utopia at work, on a grand scale and very close at hand. They had not to wait, like other evolutionary idealists, for the slow approach of something nearer to their dreams; or to leave it merely as a promise to posterity. They had not to wait for it as for a distant thing like the vision of a future state; but in the flesh they had seen their Paradise. And they were very silent for five years.

The thing died at last, and the stench of it stank to the sky. It might be thought that so terrible a savour would never altogether leave the memories of men; but men's memories are unstable things. It may be that gradually these dazed dupes will gather again together, and attempt again to believe their dreams and disbelieve their eyes. There may be some whose love of slavery is so ideal and disinterested that they are loyal to it even in its defeat. Wherever a fragment of that broken chain is found, they will be found hugging it. But there are limits set in the everlasting mercy to him who has been once deceived and a second time deceives himself. They have seen their paragons of science and organisation playing their part on land and sea; showing their love of learning at Louvain and their love of humanity at Lille. For a time at least they have believed the testimony of their senses. And if they do not believe now, neither would they believe though one rose from the dead; though all the millions who died to destroy Prussianism stood up and testified against it.

### Global Birth Control

The day will come (if the world does not before that time crash into ruin) when the higher and more intelligent communities will make sure that no danger spot, no infectious sore of over-procreation ignorance continues spawning into miserable multitudes in any quarter of the globe to risk the re-infection of the world by such an epidemic of insanity as war.—BIRTH CONTROL NEWS, JULY, 1922

### Keeping the Coloured Out

I do not see how we can hope permanently to be strong enough to keep the coloured races out; sooner or later they are bound to overflow, so the best we can do is to hope that those nations will see the wisdom of Birth Control. . . . We need a strong international authority.—BERTRAND RUSSELL, BIRTH CONTROL NEWS, DEC.1922

*EUGENICS AND OTHER EVILS*

# Hereditary Talent and Character

## *by Francis Galton*

---

**Editor:** Eugenic-like ideas are not new. Humanity may have begun some form of controlled breeding shortly after animals were domesticated, and Plato's *Republic* (4th century B.C.) described a society where those same principles were applied to people. But the modern move to link the breeding of people to science in general and to Darwinian evolution in particular began with a two-part article by Charles Darwin's cousin, Francis Galton (1822–1911) that appeared in *Macmillan's Magazine* in 1865, a mere six years after the publication of *Origin of Species*. Throughout his life Galton remained proud of that seminal article. The research technique he pioneered in there—equating high levels of "fitness" with inclusion in "who's who" directories—continued to be used in scientific research for roughly a century. What follows are extracts from that 1865 article. Note that Galton began by drawing a parallel to breeding methods used with animals. That theme constantly recurs in eugenic thinking. It is denied with equal regularity when the dehumanizing parallels to animal breeding hinder public support. In this appendix, the subtitles are my own.

---

### HUMAN TRAITS CAN BE CONTROLLED

The power of man over animal life, in producing whatever varieties of form he pleases, is enormously great. It would seem as though the physical structure of future generations was almost as plastic as clay, under the control of the breeder's will. It is my desire to show, more pointedly than—so far as I am aware—has been attempted before, that mental qualities are equally under control.

A remarkable misapprehension appears to be current as to the fact of the transmission of talent by inheritance. It is commonly asserted that the children of eminent men are stupid; that, where great power of intellect seems to have been inherited, it has descended through the mother's side; and that one son commonly runs away with the talent of a whole family. My own inquiries have led me to a diametrically opposite conclusion. I find that talent is transmitted by inheritance in a very remarkable degree; that the mother has by no means the monopoly of its transmission; and that whole families of persons of talent are more common than those in which one member only is possessed of it. I justify my conclusions by the statistics I now proceed to adduce, which I believe are amply sufficient to command conviction.

—*Macmillan's*, 12 (1865), 157.

<hr>

──────── **Editor's Note** ────────

With some humor, Chesterton described how eugenists avoid direct statements. In the passage that follows, notice that beneath the soothing words, each of these ten "deeply blushing young men" is to be put under enormous social and economic pressure to marry a particular woman selected on "established principles" and that the two are then expected to produce children who are "eminent servants of the State."

<hr>

## ENDOWING EMINENT MARRIAGES

Let us, then, give reins to our fancy, and imagine a Utopia—or a Laputa, if you will—in which a system of competitive examination for girls, as well as for youths, had been so developed as to embrace every important quality of mind and body, and where a considerable sum was yearly allotted to the endowment of such marriages as promised to yield children who would grow into eminent servants of the State. We may picture to ourselves an annual ceremony in the Utopia or Laputa, in which the Senior Trustee of the Endowment Fund would address ten deeply blushing young men, all of twenty-five years old, in the following terms:—

"Gentlemen, I have to announce the results of a public examination, conducted on established principles; which show that you occupy the foremost places in your year, in respect to those qualities of talent, character, and bodily vigour, which are proved, on the whole, to do most honour and best service to our race. An examina-

tion has also been conducted on established principles among all the young ladies in this country, who are now of the age of twenty-one, and I need hardly remind you, that this examination takes note of grace, beauty, health, good temper, accomplished housewifery, and disengaged affections, in addition to noble qualities of heart and brain. By a careful investigation of the marks you have severally obtained, and a comparison of them, always on established principles, with those obtained by the most distinguished among the young ladies, we have been enabled to select ten of their names with especial reference to your individual qualities. It appears that marriage between you and these ten ladies, according to the list I hold in my hand, would offer the probability of unusual happiness to yourselves, and, what is of paramount interest to the State, would probably result in an extraordinarily talented issue. Under these circumstances, if any or all of these marriages should be agreed upon, the Sovereign herself will give away the brides, at a high and solemn festival, six months hence, in Westminster Abbey. We, on our part, are prepared, in each case, to assign 5,000*l.* as a wedding-present, and to defray the cost of maintaining and educating your children, out of the ample funds entrusted to our disposal by the State."

If the twentieth part of the cost and pains were spent in measures for the improvement of the human race that is spent on the improvement of the breed of horses and cattle, what a galaxy of genius might we not create! We might introduce prophets and high priests of civilization into the world, as surely as we can propagate idiots by mating *crétins*. Men and women of the present day are, to those we might hope to bring into existence, what the pariah dogs of the streets of an Eastern town are to our own highly bred varieties.

—*Macmillan's*, 12 (1865), 165–66.

Hidden beneath the mathematical scenario that follows is a grim agenda. As the story begins, half the population is slated for biological extermination. Initially, the methods are indirect. Society will simply make it harder for some groups to have and rear children. It takes little skill to imagine how that might be done. High taxes, limited job opportunities, crime-ridden communities, shoddy public schools, and the culture in general could be structured to make parenthood as difficult and unrewarding as possible for these people. Eventually, groups that once made up 50 percent of the population would be reduced to a small enough minority that more ruthless methods could be employed.

## IMPROVING THE HUMAN BREED

. . . . No one, I think, can doubt, from the facts and analogies I have brought forward, that, if talented men were mated with talented women, of the same mental and physical characters as themselves, generation after generation, we might produce a highly bred human race, with no more tendency to revert to meaner ancestral types than is shown by our long-established breeds of race-horses and foxhounds.

It may be said that, even granting the validity of my arguments, it would be impossible to carry their indications into practical effect. For instance, if we divided the rising generation into two castes, A and B, of which A was selected for natural gifts, and B was the refuse, then, supposing marriage was confined within the pale of the caste to which each individual belonged, it might be objected that we should simply differentiate our race—that we should create a good and a bad caste, but we should not improve the race as a whole. I reply that this is by no means the necessary result. There remains another very important law to be brought into play. Any agency, however indirect,

that would somewhat hasten the marriages in caste A, and retard those in caste B, would result in a larger proportion of children being born to A than to B, and would end by wholly eliminating B, and replacing it by A.

Let us take a definite case, in order to give precision to our ideas. We will suppose the population to be, in the first instance, stationary; A and B to be equal in numbers; and the children of each married pair who survive to maturity to be rather more than 2 1/2 in the case of A and rather less than 1 1/2 in the case of B. This is no extravagant hypothesis. Half the population of the British Isles are born of mothers under the age of thirty years.

The result of the first generation would be that the total population would be unchanged, but that only one-third part of it would consist of the children of B. In the second generation, the descendants of B would be reduced to two-ninths of their original numbers, but the total population would begin to increase, owing to the greater predominance of the prolific caste A. At this point the law of natural selection would powerfully assist in the substitution of caste A for caste B, by pressing heavily on the minority of weakly and incapable men. . . .

I hence conclude that the improvement of the breed of mankind is no insuperable difficulty. If everybody were to agree on the improvement of the race of men being a matter of the very utmost importance, and if the theory of the hereditary transmission of qualities in men was as thoroughly understood as it is in the case of our domestic animals, I see no absurdity in supposing that, in some way or other, the improvement would be carried into effect.

—*Macmillan's*, 12 (1865), 319–20

From the very beginning, Galton understood that the scientific-only point of view he was advocating clashed with the religion. According to Judeo-Christian beliefs, it is the individual, whatever their talent, who is eternal and hence of supreme value. In the scientific world-view, the individual matters little, being no more than a tiny branch in a complex and continuously changing system. Making that system still more complex was eugenic's ultimate goal. To achieve that, individuals would have to be sacrificed. Left unsaid is the fact that this achieves nothing. The universe will eventually wind down and, with it, mankind, however highly evolved.

### BELIEF IN HUMAN UNIQUENESS NOT SCIENTIFIC

We shall therefore take an approximately correct view of the origin of our life, if we consider our own embryos to have sprung immediately from those embryos whence our parents were developed, and these from the embryos of their parents, and so on for ever. We should in this way look on the nature of mankind, and perhaps on that of the whole animated creation, as one continuous system, ever pushing out new branches in all direction, that variously interlace, and that bud into separate lives at every point of interlacement.

This simile does not at all express the popular notion of life. Most persons seem to have a vague idea that a new element, especially fashioned in heaven, and not transmitted by simple descent, is introduced into the body of every newly born infant. Such a notion is unfitted to stand upon any scientific basis with which we are acquainted.

—*Macmillan's*, 12 (1865), 322.

———— **Editor's Note** ————

Occasionally, eugenists slip into open racism, as Galton does here. However, the real evil of eugenics doesn't lie in its labeling some physically different groups more or less "fit" than others. It lies in the belief that people have different values. Once that is accepted, it matters little whether the "unfit" are targeted as groups of races and classes or simply as individuals.

### LOWER RACES REMAIN IMMATURE

Another difference, which may either be due to natural selection or to original differences of race, is the fact that savages seem incapable of progress after the first few years of their life. The average children of all races are much on a par. Occasionally, those of the lower races are more precocious than the Anglo-Saxons; as a brute beast of a few weeks old is certainly more apt and forward than a child of the same age. But, as the years go by, the higher races continue to progress, while the lower ones gradually stop. They remain children in mind, with the passions of grown men.

—*Macmillan's*, 12 (1865), 326.

———— **Editor's Note** ————

At the very close of his article, Galton again turned to religion. Both Judaism and Christianity assert that humanity has fallen from a once high position. Galton holds the contrary position and claims that we are "rapidly rising from a low one." Left unsaid is a belief that Galton and those like him represent the vanguard of this new civilization. Eugenics is anything but modest.

### SAVAGES FROM THE BEGINNING

The sense of original sin would show, according to my theory, not that man was fallen from a high estate, but that he was rapidly rising from a low one. It would therefore confirm the conclusion that has been arrived at by every independent line of ethnological research—that our forefathers were utter savages from the beginning; and, that, after myriads of years of barbarism, our race has but recently grown to be civilized and religious.

—*Macmillan's*, 12 (1865), 327.

# B

# Eugenics: Its Definition, Scope, and Aims

## *by Francis Galton*

---

**Editor:** By its very nature, eugenic science was most closely linked to evolutionary biology and to the social sciences. Far from being a pseudo-science deplored by genuine science, eugenics was commonly regarded as the first glow of a golden new age in which science would reign supreme, creating a better world and, most important of all, a better humanity. In this article, giving in its entirety, Galton addresses a prestigious group of scientists and intellectuals. (For a partial list of those present, see Appendix D.) Note the moral relativism in the first paragraph and his attempt to establish a substitute for ethics based solely on biology. Chesterton would have had great fun with the nonsense in Galton's quaint tale about talking animals. No thinking rabbit, he might remind us, "would agree that it was better" for the wolves "to be well-fitted for their part in life"—the devouring of little rabbits. Galton's blindness in this area tellingly illustrates that the eugenically inclined invariably take a top-down approach to society. That elitist mindset in turn links—in their zeal for eugenics—groups that superficially seem so different, groups ranging from scientists with social agendas to leftist intellectuals and the more ruthless sorts of capitalists. In another tale of talking animals, George Orwell captured that mindset brilliantly. In the closing scene of *Animal Farm* he has the animals look, "from pig to man, and from man to pig, and from pig to man again; but already it was impossible to say which was which." In the end, all eugenists share a common attitude. Eugenists and their ideological kin view the great bulk of the humanity as domesticated animals whose lives and breeding ought to be firmly under their control. Galton, you will note, uses exactly that analogy in this article.

---

Read before the Sociological Society at a meeting in the School of Economics (London University), on May 16, 1904. Professor Karl Pearson, F.R.S., in the chair.[1]

Eugenics is the science which deals with all influences that improve the inborn qualities of a race; also with those that develop them to the utmost advantage. The improvement of the inborn qualities, or stock, of some one human population will alone be discussed here.

What is meant by improvement? What by the syllable *eu* in "eugenics." whose English equivalent is "good"? There is considerable difference between goodness in the several qualities and in that of the character as a whole. The character depends largely on the *proportion* between qualities, whose balance may be much influenced by education. We must therefore leave morals as far as possible out of the discussion, not entangling ourselves with the almost hopeless difficulties they raise as to whether a character as a whole is good or bad. Moreover, the goodness or badness of character is not absolute, but relative to the current form of civilization. A fable will best explain what is meant. Let the scene be the zoölogical gardens in the quiet hours of the night, and suppose that, as in old fables, the animals are able to converse, and that some very wise creature who had easy access to all the cages, say a philosophic sparrow or rat, was engaged in collecting the opinions of all sorts of animals with a view of elaborating a system of absolute morality. It is needless to enlarge on the contrariety of ideals between the beasts that prey and those they prey upon, between those of the animals that have to work hard for their food and the sedentary parasites that cling to their bodies and suck their blood, and so forth. A large number of suffrages in favor of maternal affection would be obtained, but most species of fish would repudiate it, while among the voices of birds would be heard the musical protest of the cuckoo. Though no agreement could be reached as to absolute morality, the essentials of eugenics may be easily defined. All creatures would agree that it was better to be healthy than sick, vigorous than weak, well-fitted than ill-fitted for their part in life; in short, that it was better to be good rather than bad specimens of their kind, whatever that kind might be. So with men. There are a vast number of conflicting ideals, of alternative characters, of incompatible civilizations; but they are wanted to give fulness and interest to life. Society would be very dull if every man resembled the highly estimable Marcus Aurelius or Adam Bede. The aim of eugenics is to represent each class or sect by its best specimens; that done, to leave them to work out their common civilization in their own way.

A considerable list of qualities can easily be compiled that nearly everyone except "cranks" would take into account when picking out the best specimens of his class. It would include health, energy, ability, manliness, and courteous disposition. Recollect that the natural differences between dogs are highly marked in all these respects, and that men are quite as variable by nature as other animals of like species. Special aptitudes would be assessed highly by those who possessed them, as the artistic faculties by artists, fearlessness of inquiry and veracity by scientists, religious absorption by mystics, and so on. There would be self-sacrificers, self-tormentors, and other exceptional idealists; but the representatives of these would be better members of a community than the body of their electors. They would have more of those qualities that are needed in a state—more vigor, more ability, and more consistency of purpose.

---

1. Editor: This text is from: Francis Galton, "Eugenics: Its Definition, Scope, and Aims" in *The American Journal of Sociology* X, no. 1 (July 1904), 1–6.

The community might be trusted to refuse representatives of criminals, and of others whom it rates as undesirable.

Let us for a moment suppose that the practice of eugenics should hereafter raise the average quality of our nation to that of its better moiety at the present day, and consider the gain. The general tone of domestic, social, and political life would be higher. The race as a whole would be less foolish, less frivolous, less excitable, and politically more provident than now. Its demagogues who "played to the gallery" would play to a more sensible gallery than at present. We should be better fitted to fulfil our vast imperial opportunities. Lastly, men of an order of ability which is now very rare would become more frequent, because the level out of which they rose would itself have risen.

The aim of eugenics is to bring as many influences as can be reasonably employed, to cause the useful classes in the community to contribute *more* than their proportion to the next generation.

The course of procedure that lies within the functions of a learned and active society, such as the sociological may become, would be somewhat as follows:

1. Dissemination of a knowledge of the laws of heredity, so far as they are surely known, and promotion of their further study. Few seem to be aware how greatly the knowledge of what may be termed the *actuarial* side of heredity has advanced in recent years. The *average* closeness of kinship in each degree now admits of exact definition and of being treated mathematically, like birth- and death-rates, and the other topics with which actuaries are concerned.

2. Historical inquiry into the rates with which the various classes of society (classified according to civic usefulness) have contributed to the population at various times, in ancient and modern nations. There is strong reason for believing that national rise and decline is closely connected with this influence. It seems to be the tendency of high civilization to check fertility in the upper classes, through numerous causes, some of which are well known, others are inferred, and others again are wholly obscure. The latter class are apparently analogous to those which bar the fertility of most species of wild animals in zoölogical gardens. Out of the hundreds and thousands of species that have been tamed, very few indeed are fertile when their liberty is restricted and their struggles for livelihood are abolished; those which are so, and are otherwise useful to man, becoming domesticated. There is perhaps some connection between this obscure action and the disappearance of most savage races when brought into contact with high civilization, though there are other and well-known concomitant causes. But while most barbarous races disappear, some, like the negro, do not. It may therefore be expected that types of our race will be found to exist which can be highly civilized without losing fertility; nay, they may become more fertile under artificial conditions, as is the case with many domestic animals.

3. Systematic collection of facts showing the circumstances under which large and thriving families have most frequently originated; in other words, the *conditions* of eugenics. The definition of a thriving family, that will pass muster for the moment at least, is one in which the children have gained distinctly superior positions to those who were their classmates in early life. Families may be considered "large" that contain not less than three adult male children. It would be no great burden to a society including many members who had eugenics at heart, to initiate and to preserve a large collection of such records for the use of statistical students. The committee charged with the task would have to consider very carefully the form of their circular and the persons

intrusted to distribute it. They should ask only for as much useful information as could be easily, and would be readily, supplied by any member of the family appealed to. The point to be ascertained is the *status* of the two parents at the time of their marriage, whence its more or less eugenic character might have been predicted, if the larger knowledge that we now hope to obtain had then existed. Some account would be wanted of their race, profession, and residence; also of their own respective parentages, and of their brothers and sisters. Finally the reasons would be required why the children deserved to be entitled a "thriving" family. This manuscript collection might hereafter develop into a "golden book" of thriving families. The Chinese, whose customs have often much sound sense, make their honors retrospective. We might learn from them to show that respect to the parents of noteworthy children which the contributors of such valuable assets to the national wealth richly deserve. The act of systematically collecting records of thriving families would have the further advantage of familiarizing the public with the fact that eugenics had at length become a subject of serious scientific study by an energetic society.

4. Influences affecting marriage. The remarks of Lord Bacon in his essay on *Death* may appropriately be quoted here. He says with the view of minimizing its terrors: "There is no passion in the mind of men so weak but it mates and masters the fear of death. . . . . Revenge triumphs over death; love slights it; honour aspireth to it; grief flyeth to it; fear preoccupateth it." Exactly the same kind of considerations apply to marriage. The passion of love seems so overpowering that it may be thought folly to try to direct its course. But plain facts do not confirm this view. Social influences of all kinds have immense power in the end, and they are very various. If unsuitable marriages from the eugenic point of view were banned socially, or even regarded with the unreasonable disfavor which some attach to cousin-marriages, very few would be made. The multitude of marriage restrictions that have proved prohibitive among uncivilized people would require a volume to describe.

5. Persistence in setting forth the national importance of eugenics, There are three stages to be passed through: (1) It must be made familiar as an academic question, until its exact importance has been understood and accepted as a fact. (2) It must be recognized as a subject whose practical development deserves serious consideration. (3) It must be introduced into the national conscience, like a new religion. It has, indeed, strong claims to become an orthodox religious tenet of the future, for eugenics co-operate with the workings of nature by securing that humanity shall be represented by the fittest races. What nature does blindly, slowly, and ruthlessly, man may do providently, quickly, and kindly, As it lies within his power, so it becomes his duty to work in that direction. The improvement of our stock seems to me one of the highest objects that we can reasonably attempt. We are ignorant of the ultimate destinies of humanity, but feel perfectly sure that it is as noble a work to raise its level, in the sense already explained, as it would be disgraceful to abase it. I see no impossibility in eugenics becoming a religious dogma among mankind, but its details must first be worked out sedulously in the study. Overzeal leading to hasty action would do harm, by holding out expectations of a near golden age, which will certainly be falsified and cause the science to be discredited. The first and main point is to secure the general intellectual acceptance of eugenics as a hopeful and most important study. Then let its principles work into the heart of the nation, which will gradually give practical effect to them in ways that we may not wholly foresee.

Francis Galton, London

# The Progress of Eugenics

## C. W. Saleeby

**Editor:** In this book Chesterton makes a number of references to C. W. Saleeby. These quotes in this appendix are taken from Saleeby's *The Progress of Eugenics* (London: Cassell, 1914). They provide an overview of mainstream eugenics in Britain just before World War I at the time when Chesterton was writing much of this book. Note, once the author's evasions, are cast aside, how accurate Chesterton's description of eugenics was. (The headings are from Saleeby's book.)

### EUGENICS AND MARRIAGE

Thirdly, the people called eugenists do not seek the abolition of marriage. They indeed assert their intention of judging all human institutions by their supreme criterion—the quality of the human life they produce—and thus they may condemn certain aspects of marriage as we practise it. Undoubtedly the eugenist declines to accept conventional, legal, or ecclesiastical standards of judgment in this or any other matter, but inquiry compels him to recognise in marriage the foremost and most fundamental instrument of his purpose. Only it must be eugenic marriage. The Church and the State and public opinion may permit the marriage of the feeble-minded girl of sixteen, or a marriage between a diseased inebriate and a maiden clear-eyed like the dawn; but the eugenist has regard to the end thereof, and he is false to his creed if he does not declare that these are crimes and outrages perpetrated alike upon the living and the unborn. Those whom the devil hath joined together he would gladly put asunder. If this is to "attack marriage," then he does attack marriage. But this is rather to make a stand for marriage against the influences which now threaten to destroy it.

Popular misinterpreters and critics of eugenics say also that eugenists wish us all to be "forcibly married by the police," and that they want to substitute for human marriage and parenthood "the methods of the stud-farm." No one who has made the smallest contribution to or performed the slightest service for eugenics has ever made such idiotic and hideous proposals; and it is not easy to find any excuse for one or two comic philosophers who now have hold of the long-eared public, and who reiterate, year in and year out, these gross misstatements of the eugenic creed. [Pages 36–38]

——————— Editor's Note ———————

In the quote that follows, it is unlikely that alcoholics will find any comfort in the fact that Saleeby promises that they won't be killed, merely "so guarded and treated in the future that they shall not become parents at all." Following Chesterton's suggestion and putting eugenic schemes in concrete language, that means either forced sterilization or imprisoning these 'weed-like' people in an institution for most of their adult lives. And note too that the distinction between natural selection's killing and eugenics' birth prevention disappears when the eugenic method is abortion.

## "NATURAL SELECTION" AND EUGENICS

.... It is clear that if the same cause equally attacks two babies, one naturally weak and the other naturally strong, the weaker will be the more likely to die. If it could be contrived that the stronger was not also injured, there would be something to be said, on brutal and pseudo-eugenic lines, for the process. Once it is proved, however, that the weeding-out process makes far more weeds than it destroys, the argument for it is gone. And, further, the argument at best is not eugenic, Natural selection and eugenic selection may have the same effect and end, but they are fundamentally distinct in method. Natural selection is a selective death-rate, killing those less able to survive, but eugenic selection, in Professor Pearson's own admirable phrase, replaces this selective death-rate by a selective birth-rate: and no form of killing or permission of killing can be anything but a negation of the essential characteristic of eugenics. The eugenist has every right to say, and must never cease saying, that many children are born who should never have been born, or, rather, who should never have been con-

ceived. He has every right to say that the feeble-minded, and the alcoholic, and the insane, and those afflicted with venereal disease, must be so guarded and treated in future that they shall not become parents at all. But the instant he approves of the death of any who live, worthy or unworthy, he is talking not eugenics but its opposite, of which the most familiar and accurate name is murder. [Page 76–77]

——————— Editor's Note ———————

In the previous section and the one that follows we see hints that one factor driving eugenics was the enormous decline in the death rate in England during the latter half of the nineteenth century due to basic public health measures such as clean water and food inspection. While that lowered the death rate in all classes, it particularly benefited the poor living in cities. From the eugenic perspective, something now had to be done about those who weren't being killed, in proper Darwinian fashion, by polluted water.

## THE INADEQUACY OF "DARWINISM"

.... Natural selection only selects in the sense that all rejection involves selection, and that to refuse is also to choose. As such, it applies solely to what I have called negative eugenics, which seeks to limit the number of the defective and diseased. Yet even here "Darwinism" is worth less than nothing to us without qualification. In an early chapter of his *Descent of Man* Darwin pointed out that we cannot apply the principle of natural selection to the abolition of hospitals and of mercy, because we should thus lose priceless things. In his Romanes lecture ["Evolution and Ethics," 1893] Huxley stated the dilemma still more cogently, showing that "moral evolution" is opposed to "cosmic evolution" (conceived in terms of natural selection), and consists in arresting the natural process. Out of this dilemma he saw no way. But it is a dilemma only because we wrongly suppose that "cosmic evolution" depends on a merely destructive principle. When we once firmly grasp the truth that the way to maintain and magnify life is not by destruction, that no amount of killing is creation, we shall see that the valuable part of Darwin's teaching can be reconciled with morality in this case, Huxley's difficulty notwithstanding. *We can distinguish between the right to live and the right to become a Parent.* That is the principle I laid down five years ago, and I am glad to see that in a recent lecture Dr. Heron, of the Galton Laboratory, has accepted it. We can do our best for the life that is, but can follow Nature and transcend her by mercifully forbidding it to reproduce its defect. That is why I define negative eugenics as "the discouragement of unworthy *Parenthood*," a project which involves no killing, and is morally at the opposite pole from that of the purely "Darwinian" eugenists, who advocate a return to natural selection with its destruction of the unfortunate, and whom I define as the "better-dead" school of eugenists. For that is what so-called "Darwinism" leads to: the championship of infant mortality, contempt for mercy, enmity to social reform, and the prostitution of divine eugenics to the diabolical creed of "Each man for himself, and the devil take the hindmost." [Pages 150-52]

──────── Editor's Note ────────

In the first quotation from Saleeby, he had insisted that, "the people called eugenists do not seek the abolition of marriage." In the following quote he admits that his variety of eugenics has to contend with others who want to block some marriages altogether.

Again, readers should step back and ask, in proper Chestertonian fashion, "What does this really mean?" In that era, marriage without children could only mean the forced sterilization of those deemed "transmissably unworthy."

## THE DISTINCTION BETWEEN MARRIAGE AND PARENTHOOD

Yet again let us examine our terms. It is to parenthood on the part of the transmissibly unworthy that we object. Negative eugenics has no right to object to their living *or to their marrying.* This must be insisted upon. Hitherto marriage and parenthood have been regarded as synonymous or equivalent by writers on eugenics, and they have said that such and such persons must not marry, when what they meant was that these persons must not become parents. [Page 179]

──────── Editor's Note ────────

In the quote that follows we see that some of the more serious eugenists suspected that there would prove to be no clear distinction between the fit and unfit.

## THE PROBLEM OF THE "IMPURE DOMINANT"

The importance of the distinction upon which I feel compelled to insist here will become apparent when we consider the real meaning of the American demonstration that many serious defects are Mendelian recessives. It is that there are many persons in the community, personally normal, who are nevertheless "impure dominants" in the Mendelian sense, and half of whose germ-cells accordingly carry a defect. According to a recent calculation, made in one of the bulletins of the Eugenics Record Office, about one-third of the population in the United States is thus capable of conveying mental deficiency, the "insane tendency," epilepsy, or some other defect. We may hope that this estimate is far in excess of the facts, but certainly there are many such persons, and their number would be increased if Dr. Davenport's advice as to the mating of defectives with normal persons were followed, for all their offspring would then belong to this category. [Page 181]

## A GREAT LEGISLATIVE ACHIEVEMENT

. . . A second Bill is now, however, the Mental Deficiency Act, 1913, which came into force on April 1st, 1914. Thus we may record, as the greatest achievement in the progress of modern eugenics, the coming into force in its decennial year of a beneficent measure which will, for the first time, take kindly care of the mentally defective as long as they need it, and in so doing will protect the future. The permanent care for which the Act provides is, under another name, the segregation which the principles of negative eugenics require in this case. Here I am mercifully absolved from the necessity of repeating the arguments in favour of segregation, of which a decades reiteration has made me weary; so far as Great Britain is concerned the thing is, in a sense, done, though local authorities will yet need much stimulation and judicious abuse. In the United States public opinion and understanding appear to be so far advanced that the American reader need not be appealed to. [Pages 188–89]

# Two Decades of Eugenics

## *by C. W. Saleeby*

This is the published summary of a paper that Dr. Saleeby read at a meeting of the Sociological Society, Leplay House, London on May 16, 1924. It was later published in *The Sociological Review* 16 (July 1924), 251–53. (The Galton speech mentioned in the first paragraph is republished in this book as Appendix B.) Notice how Chesterton's *Eugenics and Other Evils* draws Dr. Saleeby's ire in the third paragraph. As you read, ask yourself what eugenists will be forced do with those who do not accept their scientific "new religion," but instead insist on marrying and begetting as their own religion or personal desires direct. (In this appendix, I have made bold text needing emphasis.)

Twenty years ago to-day, on May 16th, 1904, Francis Galton, a veteran student of Nature and mankind, emerged from his retirement and read before our newly formed Sociological Society a paper on "Eugenics: its Definition, Scope and Aims."[1] In printed form it is shorter than this paper. The most recent addition to my library of eugenics, twenty years afterwards, is a bibliography[2] running to 514 large pages and containing the names of contributions almost running, at a rough estimate, into six figures. Evidently my method here must be indicative and summary.

The illustrious old gentleman, so splendid in physique and face and voice, was merely stating in the fewest possible words the evident argument from his own works on hereditary genius and the inheritance of human faculty: to which he applied a purposive, human and humane version of the principle of natural selection defined by his cousin, Charles Darwin. His name and fame drew to the London School of Economics that afternoon an audience quite small in numbers, but including such men as the Chairman, Professor Karl Pearson; Dr. Maudsley, Dr.

---

1. Published in *Sociological Papers* 1904 (Leplay House, 65 Belgrave Road, S.W.), and reprinted in 1908 in *Essays in Eugenics* (Eugenics Education Society).

2. *Bibliography of Eugenics*: by Samuel J. Holmes, Professor of Zoology in the University of California. University of California Press, 1924.

Mercier, Professor Weldon, Mr. Benjamin Kidd, all now gone; and Dr. Robert Hutchison, Mr. H. G. Wells, and Professor L. T. Hobhouse. The written communications were vastly more valuable than the discussion carried on by the learned persons named; and I must at least quote the first paragraph from the contribution by **Mr. Bernard Shaw who wrote: "I agree with the paper, and go so far as to say that there is now no reasonable excuse for refusing to face the fact that nothing but a eugenic religion can save our civilisation from the fate that has overtaken all previous civilisations."**

Incomparably the most useful thing to be done here and now would be to copy out Galton's paper and let it be read, twenty years afterwards, by large numbers of serious and interested readers who have never read it all, and whose notions of eugenics are not improbably far remote from anything which the founder of eugenics would recognise or approve. "That we should be forcibly married by the police is the kind of description of eugenics which we owe, for instance, to Mr. Chesterton, and it is a fair sample of one of the many perversions and misrepresentations which the eugenic idea has suffered from that memorable May afternoon to this day.

Galton had invented the word many years previously, and in view of present Olympic contests in Paris, not to mention the humble prospects of our representatives there as compared, say, with those of the tiny Finnish nation, we may remind ourselves that the word *eugenes,* well-descended, was applied by the Greeks to, for instance, a youthful winner in the Olympic Games when they observed that his father, in his time, had been a winner also—perhaps in the same "event," or kind of contest. Galton defined eugenics as "the science which deals with all influences that improve the inborn qualities of a race; also with those that develop them to the utmost advantage." The judicious

reader will observe that in this original definition (afterwards altered under other influences much for the worse, in my judgment) Galton recognises not only "nature" but the whole of "nurture." The so-called eugenics which ignores nurture, and which has usurped the name in this country, is beneath contempt, intellectually and morally. It takes the noble name of Galton in vain and prostitutes a sublime idea to the service of selfishness and greed and jealousy of the less fortunate—the most damnable of all sentiments.

At once everybody began to talk, and nearly everybody to write, about eugenics, and we have done so ever since. To myself, a youngster just finishing an academic thesis for a doctorate, upon the influences exercised by our cities over the health of the citizens, that brief address was a revelation and an inspiration; before the meeting was over I had resolved to give up the practice of medicine in order to follow Galton. Twenty years afterward it seems to me that no other choice was possible.

But if eugenics was to be a religion, it might very well be used for that form of religion according to which "Church" looks down upon "Chapel." Here, in the unexamined assumptions of a certain pseudo-eugenics, was an excuse for snobbery, a new warrant for class hatred of the less fortunate by the more so, and an urgently needed argument against social reform. Already, shortly before this birthday of eugenics, some few of us had been inveighing against the facts, at that time most abominable, of infant mortality. To save infants and mothers might cost a little money—less than it cost to bury them, as a matter of statistical fact, we replied—and the avaricious and brutal, who, being such, should indeed have been promptly segregated or sterilised on eugenic grounds, invoked eugenics to argue that any efforts to save infants were contrary to natural selection, as if the slums were natural. Two schools of eugenists at once arose, whom Professor Patrick Geddes called respectively the Magians and the Herodians; the latter I myself, with convenient ambiguity, called the better-dead school. It still thrives, and strives to persuade the nation that open cesspools ennoble the race; but its hours are numbered and it will shortly go to its own place.

Let me recall a closing passage from Galton's paper:

**"Eugenics must be introduced into the national conscience like a new religion. It has, indeed, strong claims to become an orthodox religious tenet of the future, for it, co-operates with the workings of Nature by securing that humanity shall be represented by the fittest races.** What Nature does blindly, slowly and ruthlessly, man may do providently, quickly and kindly. As it lies within his power, so it becomes his duty to work in that direction, just as it is his duty to succour neighbours who suffer misfortune. The improvement of our stock seems to me one of the highest objects that we can reasonably attempt. We are ignorant of the ultimate destinies of humanity, but feel perfectly sure that it is as noble a work to raise its level in the sense already explained, as it would be disgraceful to abase it. I see no impossibility in Eugenics becoming a religious dogma among mankind."

Two decades afterwards, the plain question is not how much eugenics has been talked and written, but how much has been practised. The answer depends upon the particular part of eugenics which we are considering. **If it be the increase of the birth-rate amongst those stocks or families where parenthood would seem to be most desirable in the interests of the race, the answer is that we have not merely not practised eugenics, but that we have steadily practised its opposite.** The wide dissemination of the methods of contraception has served our appetites and our personal interests; we exhort each other to do as we should, and meanwhile do as we please. The reason is perfectly evident—"What has posterity done for us?"

If we are to practise Galtonian eugenics—or positive eugenics as I later called it with Galton's approval, we must have compelling motives. Remember that the problem is new; no longer does sexual appetite play a part; we know too much for that. If we are to be practising eugenists, our religion must have the force of a religion; it must determine conduct. Other religions, when and where they have worked, have provided motives which swayed the hearts and dictated the deeds of men; motives worthy or unworthy, such as gratitude and love for a saviour, hope of heaven, fear of hell. The religion of eugenics affords no motives to us of to-day to compel us to practise our creed. The parental instinct may be as strong in us as in men and women of the past, but other instincts and desires make their appeals; cats and dogs are available, with manifest advantages over children, and we can satisfy our parental instincts upon them. For myself, I see not the remotest prospect of any turn of the tide, which was already flowing in the wrong direction, and has steadily set away from Galton's noble ideal ever since he announced it twenty years ago.

——————— **Editor's Note** ———————

Saleeby was describing eugenics in a time of transition. Disdaining the use of police power, positive eugenics clearly has failed to motivate parenthood even among those 'desirables' who profess its ideals. As a result, virtually all the movement's effort will need to be devoted to curtailing the birthrate of the "unworthy." Though Saleeby does not say so, it seems rather obvious that such people will have to be forced to have few or no children since they have no reason for being interested in a 'religion' that regards them as worthless.

Far otherwise, fortunately, is, or will be, the case with those other departments of eugenic practice to which, in past years, I have given names. Some of which are now in general use. **Whilst positive eugenics—the encouragement of worthy parenthood— remains a dream, or little more, much has been done in the direction of negative eugenics—the discouragement of unworthy parenthood.** We have in this country, for instance, the Mental Deficiency Act, in which, at least, the principle of lifelong care, involving segregation of the feeble-minded, is recognised. Much also has been done in respect of preventive eugenics—the protection of parenthood from the racial poisons, such as venereal disease, lead and alcohol. Galton never considered this subject at all; it is evidently part of practical eugenics.

————— Editor's Note —————

In his closing words, Saleeby tried to bridge the gap between the children-are-like-their-parents fatalism of genetics and the more numerous rationale that environmentalism offers to those who want to manipulate and shape people after birth. But this debate between heredity and environment is pointless, since the advocates of both are simply choosing different rationale for treating people as domesticated animals who exist primarily to provide the raw material for "a nobler race."

Nor did he consider, nor has anyone yet considered as we should, the further possibility which I call constructive eugenics—the enhancement of human quality by good and well-devised nurture of future parent and parenthood from generation to generation. This, I believe, is now happening in the United States, where health and vigour steadily improve, despite the absence of anything we can properly call selection, natural or eugenic. Evidently it is conceivable that, just as inhalation of lead dust injures the germ plasm, so something else, such as right exposure to sunlight, might improve the germ plasm. Such ideas were in the minds of none who contributed to that afternoon's proceedings twenty years ago, because the biological world was then dominated by Weismannism, and a measure of influence was accredited to heredity, as against nutrition, which no serious student to-day would recognise for a moment. The germ plasm and its constituent units—they have dozens of more or less misleading names which matter nought—were regarded as existing in some private universe of their own, wholly inaccessible to ordinary mundane influences, like the Gods upon Olympus. To-day the science of experimental evolution, still only in its infancy, has disposed of those incredible phantasies. The master word is not heredity, nor infection, but nutrition, including the nutrition of the reproductive glands and their characteristic cell-products; and from the study of nutrition we shall learn how to grasp this sorry scheme of things entire which we call modern civilization, and shatter it to bits, and then remould it nearer to that world of light and air and water and food and work and play and love and worship, which the laws and heart of life desire; we may thus with Evolution—or Fate — conspire; command Nature by obeying her; and achieve constructive eugenics, building nobler individuals and through them a nobler race: if we will.

*EUGENICS AND OTHER EVILS*

E

# *Eugenics Review*

**Editor:** This appendix examines what the London-based *Eugenics Review* was saying during the period of Chesterton's book. Since eugenists often said one thing in public and another in discussions among themselves, this offers an excellent opportunity to see just how successful Chesterton was in penetrating their public mask. Note that in the first two articles the magazine pours scorn on its arch-foe Chesterton, who was well-known as a 'reactionary' Catholic. Though the fact is often left unsaid today, eugenists never had to face similar attacks from the political left. Liberals, feminists, and socialists either openly supported eugenics or remained conspicuously silent as it claimed the right to sterilized and institutionalize a population that ran, in the very least, into the tens of thousands. (The subtitles in this appendix come from *Eugenics Review*.)

## *EUGENICS AND OTHER EVILS* REVIEWED

The only interest in this book is pathological. It is a revelation of the ineptitude to which ignorance and blind prejudice may reduce an intelligent man. Mr. Chesterton's bogey is 'the scientifically organised State;' and he flatters himself that after our experience with Prussia (which, by the way, faced the world in arms and nearly beat it) 'no Englishman will ever again go nosing round the stinks of that low laboratory.' So we shall not admire 'the good Eugenist who, on his fiancée falling off a bicycle, refused to marry her.' We shall marry as we feel inclined, and 'leave the rest to God.' 'There is no reason to suppose that Dr. Karl Pearson, is a better judge of a bridegroom than the bridegroom is of a bride.' Sanity or insanity 'is only opinion.' 'Health is simply nature, and no naturalist ought to have the impudence to understand it.' 'I believe in witchcraft, and I have more respect for the old witchfinders than for the Eugenists.' These are flowerets culled at random from Mr. Chesterton's garden. It is not worth while to comment upon them, or upon the book which contains them.

M. R. Inge
—*Eugenics Review,* 14 (Apr. 1923), 53.

——— **Editor's Note** ———

The M. R. Inge of the article above could be the Wr. R. Inge of the article that follows. William Ralph Inge was an prominent eugenists and a liberal Anglican clergyman best known as the "Gloomy Dean" for his chronically pessimistic attitude. A belief that England was filling up with "degenerates" fit well with his pessimism, and the idea that science should rid society of such people meshed all too well with his liberalism.

For an example of how Inge distorted the Sermon on the Mount to defend his eugenic point of view, see the last article in Appendix H. To examine Inge's views in greater detail consult his *Outspoken Essays* (Second Series, 1922). The "stinks" that Chesterton had detected in German laboratories also drifted through Inge's mind. On page 146, he claimed: "The ridiculous dogma that men are born equal is dead if not buried. The 'sanctity of human life' must give way to the obvious truth that a garden needs weeding."

English eugenists might quibble with their German counterparts about precisely which groups constituted human 'weeds.' But there was no difference between them about what the ultimate fate 'weed-like' groups should be.

---

### FILLING ENGLAND WITH DEGENERATES

[From a review of Fallon, Valere, S. J. *Eugenics*. 1923.]

.... Professor Vallon mentions, without sympathy, the hysterical denunciations of Eugenics by Messrs. Belloc and Chesterton. We may wonder why these popular writers and journalists should wish to fill England with degenerates. But they have a reason for their incoherent rage. The realise that Science, instead of confining itself to making bad smells in laboratories, is calmly preparing to lead a social and moral revolution, a revolution in which neither medieval casuistry nor Marxian class-war will count for anything at all. The great struggle of the future will be between Science and its enemies. I can see no reason why the Christian religion should be on the side of the powers of darkness.

Wr. R. Inge

—*Eugenics Review,* 15 (*circa* Apr. 1924 to Jan. 1925), 506.

——— **Editor's Note** ———

Perhaps no other article in this appendix reveals the agenda of eugenists better than this proposal, sent in by one of the magazine's more influential readers. Notice how this proposal carries the implication that, as Chesterton suggested, policemen really would lurk about wedding ceremonies, making sure all the parties involved have the proper classifications.

As you read this, ask how practical this scheme really is. Would the classifications be kept private, given that the right to marry hinges on being in group A or B? Would a woman of 20 who was classified C, wait patiently until age 45 to marry, knowing that meant she would never have children? And would a man of that same age be interested in marrying a woman a full quarter of a century older? Last but not least, could the great bulk of society be so indoctrinated with eugenics that they would support laws that would imprison a young man for pursing an interested woman his own age?

Over time, the absurdity of this sort of scheme and the ineffectiveness of other, more politically feasible ones, would drive eugenics (after the horrors of Nazism, now hidden from public view) to support political agendas that give the illusion of personal choice while producing the proper eugenic effect.

---

### NOTES AND MEMORANDA

#### A Scheme of Eugenic Reform

An influential correspondent, who has given much thought to the subject, but who desires to remain anonymous, sends the following brief outline of what he regards as an effective scheme of Eugenic reform. He is, we gather, driven to this conclusion because he holds that sterilization is impractical and the advocacy of birth control will relatively diminish the numbers of the fit. Coming from the quarter it does, the scheme is well worthy of consideration. We certainly earnestly wish that such an examination as is proposed could be brought within the region of practical politics.

1.—Everyone to undergo a medical and psychological examination as for Life Insurance at say 18, and to be classed under A, B, or C.

A.—First Class.

B.—Bodily and Mental condition good enough for mating.

C.—Sub-normals, who should not breed.

II.—No one to marry without making known his or her classification—to those who should know it.

III.—Sub-normals not to marry unless the woman is over 45 years of age.

There remains the case of illegitimate connexions. It might be made penal for a sub-normal to form such a connexion with a woman under 45.

In this way a strong public opinion would be created.

The responsibility would be thrown upon the individual in a way that in many cases would be effective. The C. men would know that they could always marry by choosing a woman of the right age.

—*Eugenics Review*, 14 (Apr. 1923), 199.

——————— **Editor's Note** ———————

The following article illustrates that link Chesterton saw between eugenics and Friedrich Nietzsche's Master Morality—an ideology which bore its most dreadful fruits under Nazism—was willingly accepted by the other side in this debate. It also illustrates the vagueness in language that Chesterton had noted as characteristic of the thinking of many eugenists. Given the crude and ineffec-

tive birth control technology of that era, in practice Mullins' vague "shall not be at liberty to increase our burdens of tomorrow," meant that those judged "unfit" would be imprisoned for most of their adult lives in single-sex, state-run institutions. Without having been convicted of a single crime, they were to be treated as if they were dangerous criminals.

---

## EUGENICS, NIETZSCHE AND CHRISTIANITY

by Claud W. Mullins

The charge is often brought against Eugenists that their principles are in opposition to the teachings of Christianity. And I think it must be admitted that consistent Eugenists must find themselves out of sympathy with much of the work that is to-day carried on in the name of the Christian religion. They must quarrel with many of those whose lives are devoted to carrying into practice what is dictated to them by their conception of Christian duty. Much of modern religious and social work is unconsciously increasing the gravity of the problems with which future generations will have to deal. Short-sighted charity, state and private, is doing great harm in so far as it encourages the reproduction of the unfit, and there are many who believe that the exhortation "Be fruitful and multiply" must in a Christian country be held to apply to all sorts and conditions of men, regardless of economic or Eugenic stability. It is scarcely surprising, therefore, to find many Eugenists dissociating themselves from this code of public morals which they believe to be inspired by the Christian religion. Consciously or unconsciously they are forced to the position of Nietzsche, to whom Christianity seemed as a glorification and encouragement of the unfit among the human kind. Nietzsche divided the moralities of the world into two broad classes, the Master Morality and the Slave Morality. The former was the gospel of the barbaric age of Natural Selection, where the weak must die lest they hinder the rightful development of the strong. The Slave Morality was to Nietzsche typified in the teaching of Christianity, in which mankind is urged to care for the weak, whose estate is held out to be the highest. The ideal set by Nietzsche was beyond doubt a Eugenic ideal, To him the greatest of all questions was: What type of man are we tending to produce? He believed that the only way of uplifting a nation is to improve the type of its individual citizens. And Nietzsche saw around him the many influences that are working to-day to burden the fit of to-morrow. And he rushed to the conclusion that the Christian religion was fully revealed in the work of the short-sighted philanthropists of to-day, and he consequently cast aside its teaching as the very root of the Slave Morality. The aim of Nietzsche was to evolve great men in all the walks of life and to remove all restraint which would impede their evolution. And in modern life he saw that the strong and healthy were hampered by the care of the unfit and the weak. So Nietzsche blamed the Christian religion.

There are many who, with Nietzsche, believe that the greatest task of the present generation is to build up a better generation for the future; must all these, too, set themselves in opposition to Christianity? It is necessary for a Christian nation to care for the poor, to feed the hungry and to tend the sick; but does this necessarily mean that we must follow and adopt the Slave Morality? Under present conditions I submit that we are doing so, and the result, if we continue on our present path, must be what Nietzsche predicted, the ultimate triumph of degeneracy, which, of course, means national and racial decay and the victory of a more virile civilisation. But does not the science of Eugenics teach us how we can retain what is noble and valuable in the Slave Morality and at the same time retain the ideals of the Master Morality? As Christians we cannot return to the practice of killing off the weak and allowing the poor and the diseased to die; we cannot return to the barbarous age of Natural Selection. But surely, as Christians, we can insist that the pauper degenerate, for whom we care to-day, shall not be at liberty to increase our burdens of to-morrow. Under the Christian code of morals, as under any other, we are perfectly entitled to look forward and to see that the next generation shall not be burdened by our charity of to-day. As Christians we must be charitable to the living, but we must also have thought for the future.

I submit that there is nothing anti-Christian in the doctrines of Eugenics. Their application will shock many present-day conventions and beliefs, and they will possibly be denounced as opposed to Christian morality; but those who believe in Eugenic reforms can be inspired by the Christian ideal just as much as those who preach and practise modem charity. Eugenics loyally follows what is fundamental and true in the Christian code, though it shakes off many of the mistaken conceptions of Christian duty which are so widespread to-day. At the same time Eugenists incorporate with their practical Christianity much of what is healthy and stimulating in the warnings of Nietzsche.

—*Eugenics Review,* 4 (*circa* January, 1913), 394–95.

——————— **Editor's Note** ———————

After the horrors of Nazism became widely known, a great deal of effort was devoted to creating the impression that eugenics (and its close kin, racism) was a fringe phenomena that had never been accepted by the well-educated. As this news story illustrates, the very opposite was true, a disturbing 61 percent of Britain's future intellectual elite approved of its principles.

Thirty-three years later, in an Autumn 1946 article in *Modern Quarterly,* the J.

B. S. Haldane mentioned in this article would attack the anti-eugenic ideas in C. S. Lewis' 1946 science fiction novel, *That Hideous Strength*. Lewis' reply would come in a posthumous essay entitled "A Reply to Prof. Haldane." So, in a very real sense, after Chesterton's death in 1936 Lewis took up his anti-eugenic banner, though not in as great a detail or quite as aggressively.

## THE DEBATE ON EUGENICS AT THE OXFORD UNION

On November 13th a motion that "this house approves of the principles of eugenics" was carried by a majority of 39 votes, having 105 supporters and 66 opponents. But from the three reports of the meeting which we have been able to collect it is not possible to obtain any idea of the course of the debate. Our venerable contemporaries, The *Oxford Magazine* and *The Isis,* concern themselves only with the manner of the speakers and fill a column or a couple of columns with criticisms which are mildly patronising or laboriously smart. The younger *Varsity* has been brought up in the same tradition, but its comments show a livelier wit, and it does make fragmentary allusions to the substance of the speeches. We are told, for instance, that Mr. J. B. S. Haldane (New College), whom all three papers agreed to praise, "has no fear of eugenics thwarting him of his lady-love. For a eugenic girl is a healthy girl, and a healthy girl is an attractive girl." Eugenics has on many occasions come before debating societies in various colleges, but its inclusion among the few non-political subjects discussed by the union is of special significance as an indication that it has grown largely in esteem during the last few years as a serious topic.

—*Eugenics Review*, 5 (Nov. 1913), 385–86.

————— **Editor's Note** —————

The uncertainty of precise dates for many of these articles is a result of issues being bound without covers. This has its amusing side. The eugenists often claimed to have a greater-than-normal intelligence and foresight. But for all their claims, they did not display enough foresight to include the date of each issue within the journal itself.

A P P E N D I X

# F

# *Eugenics Review* and the Mental Deficiency Act

---

**Editor:** Chesterton directed much of his criticism of eugenics toward what was at that time its most dangerous application, the Mental Deficiency Act of 1913. This appendix looks how that act was covered in *Eugenics Review* during the time it was being debated and after it had passed. The first article is part of the published version of a lecture that Leonard Darwin (grandson of Charles Darwin) gave to the Cambridge University Eugenics Society in February of 1912. Note his plan to use public schools and social agencies to track down the "naturally unfit." Note too his admission that the real agenda of eugenists was far more coercive than the public of that day would tolerate. (I have made text bold for emphasis.)

---

### FIRST STEPS TOWARDS EUGENIC REFORM
#### *by Major Leonard Darwin*

In order to carry out any such reform effectually it is obviously necessary first to devise some method of sorting out the naturally unfit. With this object in view, the first step to be taken ought to be to establish some system by which all children at school reported by their instructors to be specially stupid, all juvenile offenders awaiting trial, all ins-and-outs at workhouses, and all convicted prisoners should be examined by trained experts in mental defects in order to place on a register the names of all those thus ascertained to be definitely abnormal. In this examination both physical and psychological tests should if possible be included, in which case the reports thus obtained would afford a good foundation for selecting out the most unfit. From the Eugenic standpoint this method would no doubt be insufficient, for the defects of relatives are only second in importance to the defects of the individuals themselves—indeed, in some cases they are of far greater importance. **Hence it is to be hoped that in the more enlightened future, a system will also be established for the examination of the family history of all those placed on the register as being unquestionably mentally abnormal, especially as regards the criminality, insanity, ill-**

health and pauperism of their relatives, **and not omitting to note cases of marked ability.** If all this were done it can hardly be doubted that many strains would be discovered which no one could deny ought to be made to die out in the interest of the nation; and in this way the necessity for legislation, such as that proposed by the Royal Commission on the care and control of the feeble-minded, would be further emphasised. For the present it would, however, perhaps be wise to confine our efforts to endeavouring to obtain an effective examination of the individuals themselves, lest our desire to look into their family histories should brand us as being too scientific for practical purposes!

The real practical question as to how to select the individuals who should be segregated—how actually to draw the line—has, it may truly be said, still been shirked in this discussion. This is no doubt true. But it must be remembered that in many similar cases in the practical affairs of life no rules indicating exactly where the line should be drawn can be laid down in words even where the most vital decisions have to be made. This is true, for instance, in many respects in deciding whether or not a man is a lunatic, an idiot, or a criminal. The answer given to such questions must in reality always depend on the judgment of men merely guided by the knowledge of broad and general principles; and in drawing the line with regard to the segregation of the Feeble-Minded or of any other class of the eugenically unfit in the interests of posterity without laying down any hard and fast rules, it can merely be said that such a proceeding would form no exception to the methods generally adopted in such cases. Under an all-wise government these guiding principles might perhaps be such as would result in the net to catch the mentally defective being spread very widely. **But it is quite certain that no existing democratic government would go as far as we Eugen-**ists **think right in the direction of limiting the liberty of the subject for the sake of the racial qualities of future generations.** It is here that we find the practical limitation to the possibility of immediate reform; for it is unwise to endeavour to push legislation beyond the bounds set by public opinion because of the dangerous reaction which would probably result from neglecting to pay attention to the prejudices of the electorate. In existing social conditions the possibility of making progress in matters of wide interest is limited by the sentiments of the nation; and it follows that one of the first steps towards Eugenic reform must be the education of the public, an end to which our efforts should therefore now be directed.

Possibly there may be many who feel as regards the foregoing discussion that, even though the views we Eugenists hold may be sound enough, yet we have been traversing a region for the present, at all events, in reality far removed from the practical affairs of life. It is to be hoped that the introduction of a Bill into Parliament in the coming session, in which the segregation of the feeble-minded will be dealt with on Eugenic principles, will dispel this illusion.

—*Eugenics Review,* 4 (*circa* April 1912), 34–35.

## THE FEEBLE-MINDED CONTROL BILL

Mr. Gresham Stewart, who drew the eleventh place in the Ballot for Private Members' Bills at the opening of the Session, has undertaken to introduce a Bill "for the better control of the Feeble-minded."

A joint meeting was held at the House of Commons of the supporters of the Feeble-Minded Control Bill and of the Mental Defect Bill. Although it had been definitely promised by a Cabinet Minister that a measure dealing with this subject should be introduced by the Government during the present session, it was nevertheless decided that a Private Member's measure should be brought forward. In the event of other business 'crowding out' the Government Bill, the passing of such a Bill was held to be desirable in order to give powers of certification and detention in the immediate future. There has been some confusion in the country at large owing to two Bills having been brought forward— the Mental Defect Bill, prepared by the Charity Organisation Committee, following closely the lines of the Recommendations of the Royal Commission, and entailing the expenditure of a considerable sum of public money, and the Feeble-Minded Control Bill drafted by the National Association for the After Care of the Feeble-Minded and this Society, which omits all mention of methods of administration, and merely legalises under suitable safeguards, the certification and detention of the Feeble-Minded.

Every student of the problem knows that a complete measure such as the Mental Defect Bill, is a necessity for the welfare of the Country, and the Eugenics Education Society heartily supports it. But as the Standing Orders of the House impose great difficulties on private members when introducing measures involving the expenditure of public funds, it was thought better to press forward the Feeble-Minded Control Bill with the view of introducing the principle of segregation, and in the belief that it would necessitate the provision of more complete administrative machinery in the near future. It was, in fact, decided to drop the larger measure for the present, and all parties in the House have combined in support of the Feeble-Minded Control Bill. It is backed by Lord Claud Hamilton, Sir George Younger, Mr. Pike Pease, Sir Charles Nicholson, Mr. Walter Rea, Mr. William Pearce, Mr. Jowett, Mr. Crookes, Mr. Pollock, Dr. Addison, and Mr. Dickenson. From this it will be seen that the measure is strictly non-party, and non-controversial. **The prospect of some real advance towards the Permanent Control of this unfortunate group in the community is at last within sight.**
—*Eugenics Review* 4 (*circa* Apr. 1912–Jan. 1913), 108–09.

## MENTAL DEFICIENCY BILL.
It is with the deepest regret that we have had to relinquish all hope of seeing this much-needed measure become law this Session. The most important clauses were discussed in Committee, and considerably amended; and a promise has been given that a Bill on the lines thus laid down will be introduced next Session. Our efforts to secure this result must not, however, be in the slightest degree relaxed; for the danger of measures being dropped if no party capital can be made out of their enactment seems to be increasing. Great numbers of Boards of Guardians have passed resolutions of protest against the dropping of the Bill, and these afford the best possible indication of the widespread demand for this measure. Members of Eugenic Societies should continue to urge on their representatives in Parliament by every available means the necessity of dealing with measures affecting the nation as a whole, and should unsparingly condemn their abandonment on account of the mere demands of party.
—*Eugenics Review* 4 (*circa* Jan. 1913), 420.

——— **Editor's Note** ———
The push for eugenic legislation that began in 1912, continued during 1913. In the following article, notice that eugenicists were supporting legislation that would have "prohibited marriage with a defective" just as Chesterton had warned. Note too their claim that the primary opposition to their cruel measures came from "extreme individualists." Eugenists seem to have felt free to make concessions to such people because they intended to interpret individual interests in such a eugenic way that the distinction between individual and community interest was of no

importance to them. Those who think that those targeted by the bill were few should note that the definition of "defective" included those "without visible means of support." In an economic depression—such as that during the 1930s—this definition would include tens of millions of British citizens. Last of all, note the stilted writing style that Chesterton delighted in mimicking.

## THE MENTAL DEFICIENCY BILL
### R. Langdon-Down

The essential difference between the Bill of this year and that of last is that the criterion that appears to have been taken by which each provision of the measure is to be justified is the ultimate interest of the individual defective. That this principle is to guide those entrusted with the administration of the Act is explicitly laid down in several places.

Special provisions as to which it might be open to question whether they were primarily in the interest of the individual have been omitted, as for example **clause 50 in the earlier Bill, which prohibited marriage with a defective.** There is, however, an amendment proposed to reinstate this clause in Committee, but the matter has not yet been reached. Nevertheless, the powers that are desirable in the interests of the individual defective are so similar to those which have been asked for in the interests of the community from a Eugenic point of view that it was hardly possible that any measure adequately meeting the former requirements could fail at the same time to safeguard the community from the multiplication of defectives. Hence the Eugenics Education Society can heartily welcome the Bill now before Parliament. . . .

For the rest the Bill has been rearranged and for the most part simplified. Less has been left to the regulations, and throughout one can trace evidence of an endeavour to conciliate the opposition of the extreme individualists who so effectively delayed the proceedings in Committee last year. . . .

Turning to the amendments made in Committee, the first is of some importance. **The first category of defectives subject to be dealt with, viz., those who are found neglected, abandoned, or cruelly treated, has been extended to include those who are "without visible means of support"; an obvious improvement from all standpoints. . . .**

Looking back, one may say that the key of the position aimed at by the Eugenics Education Society was won last year after the ground had been prepared and searched by the Feeble-minded Persons Control Bill. . . .

There is now every prospect of a measure being placed on the Statute Book which, subject to the exiguous financial provision made, will enable and require local authorities to make suitable provision for the defectives who are not already properly cared for, and who, inevitably from the nature of their condition, come to grief themselves, and are a source of untold injury to the community and to the race.

—*Eugenics Review,* 5 (*circa* Apr. 1913–Jan. 1914), 166–67.

——————— **Editor's Note** ———————

As the next article points out, attempts were made to include in the Mental Deficiency Act provisions that would have not only banned the future marriage of "defectives," but declared marriages already formed "null and void"— an act of particular cruelty. Even the amendment to protect "defectives" from sexually exploitation was probably intended to keep them from having children outside the marriages they were not be permitted to have.

## THE MENTAL DEFICIENCY ACT, 1913

When reference was last made to this matter the Standing Committee had reached Clause 63, and it was anticipated that the

Bill would safely reach the Statute Book before the close of the session. It is satisfactory to record that this result was duly achieved on August 15th. The Bill was reported to the House on July 15th, and was read a third time on July 29th. For this we are not a little indebted to the genuine and whole-hearted interest shown by Mr. McKenna in the Bill, and the determination with which he pushed it through all its stages when other measures were being sacrificed. This fate unfortunately befell the Defective and Epileptic Children Act Amendment Bill. It is important that this Bill should be pressed in [the] next session in order to fill up what would be a serious hiatus in the scheme as a whole. The amendments made in the later stages were not of a fundamental character. The clause prohibiting marriage with a defective, which appeared in the 1912 Bill, was not reintroduced, **nor was the new clause put down by Dr. Chapple to treat a marriage with a defective as null and void proceeded with.** Dr. Chapple did, however, succeed in adding an important clause (56) protecting defectives from acts of sexual immorality, procuration, etc. The Act does not come into operation until April 1st, 1914. In the meantime the Board of Control provided under the Act will have to be constituted, and the regulations governing its detailed administration will have to be drafted and laid before Parliament for thirty days. Local authorities are already busy considering the steps they propose to take to carry out the provisions of this long desired measure. For example, a sum of 12,000 pounds was voted for this purpose at a recent meeting of the Devon County Council.

—*Eugenics Review*, 5 (*circa* Apr. 1913–Jan. 1914), 290.

——— **Editor's Note** ———

Driven by the possibility of increased funding and new career opportunities, many social work professionals were zealous eugenic supporters.

## THE MENTAL DEFICIENCY ACT

The Mental Deficiency Act, which comes into force on April 1st, may be regarded as a very useful instalment of legislation. It is, perhaps, the only piece of English social law extant, in which the influence of heredity has been treated as a practical factor in determining its provisions.

**The Act does not go as far as some of its promoters may have wished, yet most good things grow slowly, and legislators were well advised, in this instance, in adopting cautious measures, where so much is debatable, so much untried, or still in experimental stages.** Much of its general utility even now depends upon two factors, the passing of the Elementary Education (Defective and Epileptic Children) Bill, introduced by Mr. Joseph Pease this Session, and the increase of the Treasury grant. Without the clauses making it compulsory to establish special schools, children above school age needing institutional care or guardianship will be left to drift into danger, save in the unlikely event of the parents of the very worst cases notifying the local authority of the necessity for the certification and segregation of their defective children. Without an increased Treasury grant, the clause in the Act (30 [i]), which makes it non-obligatory for a local authority to provide either institutions, or guardianship, for persons within their area, whom they ascertain to be mentally defective, if the money provided by Parliament is less than half the net amount of the cost, will affect more areas than is realised, as the Treasury grant is at present only 150,000 pounds. It is permissive, however, for any local authority to act without the grant, a fact which should be borne in mind by those anxious to promote the movement. So far the most active authorities are the Poor Law Guardians. Almost every union in the country is discussing the possibilities of institutional treatment.

*EUGENICS AND OTHER EVILS*

In addition to the existing colonies connected with the London and Birmingham Boards of Guardians, twenty-four unions in the North of England have already bought land for this purpose; others have their plans almost completed. Many county councils, on the other hand, have not even constituted their "committees for the care of defectives," yet there appears already, as at Wolverhampton, some little danger of disagreement between these rival authorities as to their relative spheres of action. This spirit is to be deprecated, as there seems more than enough for all to do in carrying out the provisions of the Act. The formation of the Local Committees, the maintenance in institutions or provision of guardianship for defectives, whether under Poor Law or otherwise, the establishment of special schools, and the co-ordination of voluntary effort will require the devotion and interest of every agency available for the care of defectives for some time to come.—A. H. P. KIRBY

—*Eugenics Review*, 6. (Apr. 1914), 52–53.

## MENDELISM AND FEEBLEMINDNESS
### *R. Douglas Laurie*

*Hidden Feeblemindness,* by E. M. East. The author, starting from Goddard's conclusion that feeblemindedness is transmitted as a Mendelian recessive, and from the estimate that 3 per 1,000 of the population of the United States are feebleminded, calculates that "one person out of every 14 carries the basis of serious mental defectiveness in one-half of his or her reproductive cells," and that this understates rather than overstates the facts. He suggests that investigations should be carried out with the object of determining whether the dominance of normal over defective mentality is incomplete, because should this be so, it might become possible to recognise carriers of mental defects. (*Journal of Heredity,* May, 1917; pp. 215–217.)

—*Eugenics Review*, 9 (Apr. 1917–Jan. 1918), 263.

——— **Editor's Note** ———

The article that follows illustrates the accuracy of Chesterton's warning that the Act was so vague it could be interpreted in all sorts of ways.

## THE MENTAL DEFICIENCY ACT AND ITS ADMINISTRATION
### *Evelyn Fox*

In April, 1914 the Mental Deficiency Act came into operation and after nearly four years it is possible to form some estimate of the value of the measure, of its shortcomings, and of the means available for coping with them.

To all those who remember the Committee stage of the Act, there must remain a vivid recollection of the various opposing forces: the "vested interests" in defectives which nearly wrecked the Bill. The amendments necessitated by the opposition of the Poor Law, the Education Authorities, the defenders of the "liberty of the subject," resulting in the passing of a measure peculiarly complicated, even for an Act of Parliament, and full not only of technical difficulties, but of possibilities of friction between local governing bodies. . . .

**Experience has shown that there has been a marked diversity in the standards of certifying officers, even in the area of one local authority, and even greater variation in different areas. . . .**

—*Eugenics Review,* 10 (*circa* Apr. 1918), 1.

APPENDIX

# G

# *Birth Control News* and Forced Sterilization

**Editor:** *Birth Control News* was published monthly by the London-based Society for Constructive Birth Control and Racial Progress (or C.B.C) led by a prominent feminist, Dr. Marie Stopes (1880–1958). (Her doctorate wasn't in medicine. It came from studying fossilized plants in Germany.) The C.B.C. represented the "mainstream" of the British birth control movement and was the counterpart to Margaret Sanger's American Birth Control League with its *Birth Control Review*. In this chapter I give some articles that *Birth Control News* published on state-mandated sterilization during the period when Chesterton's book was published. I have attempted to convey the emotional flavor of the original by replicating the stress in the headlines and text. (In this appendix, text bolded was bolded in the original.) When reading, readers should remember that the barrier-type birth control techniques of that day were less effective and more difficult to use than modern, hormonal contraceptives. This meant that, from the eugenic perspective, sterilization was far more important than it is today, when long-term, injectable contraceptives (and legalized abortion) can coerce sterility as surely as surgical sterilization did during the 1920s. Readers will notice that the female-led birth control movement was far more radical in its demand that society rid itself of the unfit than the male-dominated eugenics movement. Though true, this runs counter to the oft-repeated claim that the birth control movement was intended to empower women but was corrupted by eugenics. As you will see in this appendix and the next, the opposite was true. The most zealous advocates of squelching the reproductive powers of poorer women were affluent, well-educated women. In Chapter 5, Chesterton blasted the idea of "solidarity" among workers. The same can be said of women.

## WHY STERILIZE THE FIT?

### STERILIZATION OF UNFIT RAISES A HORNET'S NEST

### But No One Worries at All About the Daily Sterilization

#### NOW GOING ON OF THE FIT

An unusual excitement has been caused by the very sensible and outspoken articles published recently by the *Morning Post,* in which it was seriously discussed whether we should sterilize mental defectives and unfit persons, so as to prevent them breeding at the cost of the State.

**Howls and outcries were at once made about the "barbarous" suggestion, but not one word has been raised about the actual daily fact of the compulsory sterilization of many of our most fit.**

Young married men of the professional classes are to-day often forced by conditions to remain sterile, though they passionately desire the healthy children they could have if they did not have hordes of **defectives** to support in one way or another.

If is no good howling against sterilization—**we have got it.** Is it to be used so as to cut off the line of the fit or the unfit?

—*Birth Control News,* July 1922, 1.

#### ———— Editor's Note ————

Political activists often use a tactic called, "they're doing it elsewhere." In the U.S., Margaret Sanger would point to birth control clinics in Holland to justify her own controversial clinics. In the next article, British birth controllers pointed to sterilization laws in the U.S. as a model for what they wanted to see in their own country.

Note the early adoption of laws in Washington and California, two states considered 'progressive" and the total absence of eugenic sterilization laws in the South. In effort to deflect attention away from the fact the coerced sterilization was a progressive cause, some tele-vision documentaries have used the fact that *Buck v. Bell*—the 1927 Supreme Court case that declared forced sterilization constitutional—involved a Virginia law. But the case arose because eugenists were having a difficult time passing laws in the South or retaining them in the face of widespread and often religiously based public opposition. *Buck v. Bell* was contrived to force sterilization on those unwilling states.

---

### STERILIZATION BY LAW IN THE U.S.A.

Those who argue about sterilization in this country frequently do so as though the legal enactment of laws for sterilization was a novelty. The idea may be a novelty to the insular British; it is no novelty across the Atlantic, where for years past many States have enacted operative sterilization laws.

#### Fifteen States in the U.S.A

STERILIZATION LAWS ENACTED BEFORE 1920 IN THE FOLLOWING STATES

Indiana. . . . . . . . . . . . . . . . . . . . . 1907
Washington. . . . . . . . . . . . . . . . . 1909
California . . . . . . . . . . . . . . . . . . 1909
California . . . . . . . . . . . . . . . . . . 1913
California . . . . . . . . . . . . . . . . . . 1917
Connecticut. . . . . . . . . . . . . . . . . 1909
Nevada . . . . . . . . . . . . . . . . . . . . 1911
Iowa—
      First Statute. . . . . . . . . . . . . 1911
      Second Statute. . . . . . . . . . . 1913
      Third Statute,. . . . . . . . . . . . 1915
New Jersey . . . . . . . . . . . . . . . . . 1911
New York . . . . . . . . . . . . . . . . . . 1912
North Dakota. . . . . . . . . . . . . . . . 1913
Michigan. . . . . . . . . . . . . . . . . . . 1913
Kansas—
      First Statute. . . . . . . . . . . . . 1913
      Second Statute, . . . . . . . . . . 1917
Wisconsin . . . . . . . . . . . . . . . . . . 1913
Nebraska. . . . . . . . . . . . . . . . . . . 1915
Oregon. . . . . . . . . . . . . . . . . . . . . 1917
South Dakota. . . . . . . . . . . . . . . . 1917

The following is taken from the summary of Dr. Laughlin's forthcoming book, an abstract of which was published in *Social Hygiene*, vol. 6:—

**Sample of Data of Law from One State**
INDIANA U.S.A.

*Date of Approval of Statute*—March 9, 1907

*Reference in State Laws*—Chapter 215, Laws of 1907

*Persons Subject*—Inmates of all State institutions deemed by commission of three surgeons to be unimprovable, physically and mentally, and unfit for propagation.

*Executive Agents Provided*—Committee of experts, consisting of two skilled surgeons of recognized ability, who shall act in conjunction with regular institution physician and a board of managers of the institution.

*Basis of Selection: Procedure*—Inadvisability of procreation and improbability of improvement of mental and physical condition, in judgment of committee of experts and board of managers of the institution.

*Type of Operation Authorized*— "Such operation for the prevention of procreation as shall be decided safest and most effective."

*State's Motive*—Purely eugenic.

*Appropriations Available for Enforcing the Act*—In no case shall the consultation fee be more than $3.00 to each expert, to be paid out of the funds appropriated for the maintenance of the institution.

Similar details, which vary somewhat State by State, are available for all the other states with such laws.

—*Birth Control News,* Oct. 1922, 4.

————— **Editor's Note** —————

Eugenists were often vague about what they meant by the "unfit." The following article illustrates that epilepsy (whatever its cause) was considered justification for sterilization. Note in the articles that follow, the zeal some judges displayed for forced sterilization. It is simply not historically accurate to say that the judicial system protects us from evils inherent in other branches of government. No branch of government has a very impressive record.

---

## "BREEDING DISEASE AND CRIME"

### Judge on Sterilizing the Unfit

Before Mr. Justice Roche, at the Central Criminal Court recently, Charles Edmund Seymour, 27, tailor, pleaded "Guilty" to wounding Mrs. Adeline Bles in Hyde Park with intent to do grievous bodily harm. Mr. Eustace Fulton prosecuted; Mr. A. B. Lucy defended.

Mr. Justice Roche remarked that some time it might be a part of the English law to sterilize people with such tendencies as the prisoner, and the sooner English doctors studied the question the better and the sooner we were likely to have a different type of people to deal with.

In reply to Mr. Lucy, Dr. East, medical officer at Brixton Prison, said he had formed the opinion that the prisoner was undoubtedly a genuine epileptic. Epilepsy would make him very impulsive and lose control of himself.

Mr. Lucy said the prisoner had been incapacitated from his youth by fits. It was a pity he could not stay in a hospital, as he was a danger to society.

Mr. Justice Roche, in passing sentence, said he pitied the prisoner because he pitied epileptics and people with infirmities of that sort.

Turning to the jury, Mr. Justice Roche said: "In my judgment, the medical profession of this country would be performing a public service if they studied earnestly the question of the feasibility of sterilizing both men and women with tendencies such as the man before me has. To allow them to produce is breeding from the worst of all stock, and propagating disease and crime. I am expressing no opinion whether it is feasible or whether Parliament should pass such a measure. That depends on the examination of skilled persons as to the feasibility and risks attending. —*The Times,* October 17, 1922

This was followed by a letter to the *Times* on Sterilization Laws by Dr. M. Stopes:—

"May I point out that the sterilization by law, although perhaps a novel ideal to the insular Briton, has been in existence in the other great English-speaking nation for a long time? Fifteen States in the U.S.A. enacted sterilization laws before the year 1920. The knowledge that others have taken this important national step may, perhaps, make it easier for English men and women to consider the subject freed from that shrinking fear induced by anything that is too novel."

—*Birth Control News,* Nov. 1922, 1.

## THE INSTITUTE OF HYGIENE

Mr. Harold Cox opened the discussion of the multiplication of the unfit, advocating sterilization by the state of such forms of unfitness as feeble-mindedness. Mr. Cox also advocated the reduction of the general birth-rate.

—*Birth Control News,* Nov. 1922, 4.

——— **Editor's Note** ———

From a eugenic perspective, it does little good to sterilize someone after they have parented several "socially inadequate" children. Hence, hidden in this article, is the suggestion that "delinquent" teenage boys and "wayward" teenage girls—as determined by "experts"—should be surgically sterilized while still young.

## COURT OF CHICAGO FOR STERILIZATION

From Chicago we learn that "Sterilization of men and women who may be the parents of 'socially inadequate' children, as determined by experts, is advocated in a

volume issued by the psychopathic laboratory of the Municipal Court of Chicago."

A model law to accomplish this, which Chief Justice Olson announced, will be presented to the Illinois General Assembly.

The "Socially Inadequate Classes" are defined as:—

(1) Feeble-minded; (2) insane (including the psychopathic; (3) criminalistic (including the delinquent and wayward); (4) epileptic; (5) inebriate (including drug habitues).

—*Birth Control News,* Feb. 1923, 1.

## GYNAECOLOGIST DEMANDS STERILIZATION OF DEFECTIVES

Dr. R. A. Gibbons, Gynaecologist to the Grosvenor Hospital for Women, speaking at the Westminister and Holborn Division of the British Medical Association after their dinner at the Criterion Restaurant, urged the Sterilization of the Unfit and State Certificates of Marriage.

He spoke of the numbers of insane under our care, saying as "there is an annual increase in these numbers, it is clear that there is urgent need to do all in our power to ameliorate this state of affairs." "We should at least endeavour to secure the advent of only healthy children . . . and prevent the propagation of the mentally defective.

Dr. B. Dunlop agreed, but also put in a claim for the value of birth control to achieve this end.

—*Birth Control News,* May 1923, 1.

## MENACE TO RACE

### Doctor on Increase of Imbecility

Sterilization of affected persons was one of the measures suggested by Dr. S. Bargrave Wyborn, a retired surgeon, at a meeting of Monmouth Guardians' yesterday, as a means of checking the alarming increase of imbecility and epilepsy.

The hereditary taint, said Dr. Wyborn, had become a menace, and inter-marriages also contributed to the spread of lunacy.

Certain districts were affected in a peculiar manner.

The Board decided to urge the Ministry of Health to undertake research work and propaganda, and to consider in hereditary cases even the sterilization of the subject.

—*Birth Control News,* Aug. 1923, 1.

## STERILIZATION OF DEFECTIVES RECOMMENDED BY LAWYERS

### Vancouver Bar Association Committee Would Revise Criminal Code

*Vancouver* [British Columbia] Recommending a complete revision of the Criminal Code, a committee of lawyers has presented a report on the administration of criminal justice to the executive of the Vancouver Bar Association. . . .

Following are some of the outstanding recommendations of the committee: . . .

Sterilization of certain types of diseased and mentally defective prisoners.

—*Birth Control News,* Sept. 1923, 1.

——— **Editor's Note** ———

Perhaps the most revealing of all the articles in this series is the one that follows. Birth control was obviously spreading from the top of society downward, but birth controllers could not be certain that their chief target—the working poor—would ever adopt it or use it very enthusiastically. Eugenic critics of birth control (such as S. J. Holmes) used this as a powerful argument *against* birth control.

But, as Prof. MacBride pointed out, birth controllers had an equally powerful counter argument. They could hope for a day when there was a critical mass of people who felt that, for various reasons, they had to limit the size of their families. Those people could then be manipulated into supporting forced sterilization.

There's a further factor that goes unmentioned here. That much-longed-for day could never come as long as there remained within society groups that were educated and articulate opponents of state-control over reproduction. Hence the fury with which eugenists, birth controllers and the like attack and attempt to slander their opponents, primarily religious "reactionaries" such as Chesterton. (For an example of how birth controllers attacked him, see Appendix I.)

---

### EVOLUTION AND EUGENICS

Prof. E. W. MacBride, in a review of Prof. S. J. Holmes book *Studies in Evolution and Eugenics,* in *Nature,* writes as follows:—

". . . . Professor Holmes objects to *birth control,* because only the prudent would practise it: this is true, but what he fails to recognize is that many of the prudent and intellectual classes use birth control now; so far as they are concerned the mischief is already done. The advocates of birth control desire to teach it to the working-class, the best of whom are only too desirous of learning the means of employing it. If this end were accomplished, the lower stratum of utterly reckless and vicious people would no doubt be unaffected, but a strong public opinion would have been created which would support the application to these people of the only remedy possible (namely, sterilization)."

—*Birth Control News,* June 1924, 4.

——— **Editor's Note** ———

Forced sterilization did have its critics. The roots of birth controller hostility to Roman Catholicism lies in Chesterton's day in the protection the church and its members offered to "defectives."

---

## STERILIZATION OF DEFECTIVES DENOUNCED BY ROMAN CATHOLIC DOCTORS

Lieut.-Col. O'Gorman, C.M.G., M.D., read a violent attack on sterilization to the meeting of the St. Luke's Medical Guild. "Man," he concluded, "has no mere mutilative power over his body, and cannot delegate that power."

He pointed out that sterilization "would only bring about the same issue as artificial birth control, and thus enable a marriage to be degraded to the depths of a domesticated brothelism."

*—Birth Control News,* Sept. 1923, 1.

———— **Editor's Note** ————

There is no better indication of just how poorly the mainstream media has handled eugenics than a poll *The Guardian,* a prominent British newspaper, took in 1999 to establish who its readers considered "The Woman of the Millennium." Coming in first place was none other than Marie Stopes. One reader, in a triumph of prejudice over reality, praised Stopes for "promoting family planning and giving women control over their fertility"—an agenda that could not be more diametrically opposed to the cause to which Stopes dedicated her life.

There's also an amusing aside to Stopes' biography which suggests that the intelligence that eugenists so highly valued did not necessarily imply ordinary good sense. According to the internet web site run by the Marie Stopes International (http://www.mariestopes.org.uk/) there was a very practical reason why Stopes entered the birth control movement:

"But Marie Stopes might never have got involved in family planning if she hadn't had a disastrous marriage to fellow scientist Reginald Ruggles Gates. They had a whirlwind romance, but the relationship was close to break-down within a year. Although she was highly intelligent, it only gradually became apparent to Marie that her sex life was not quite right."

"After studying medical books in various languages in the British Library she realised her husband was impotent and that she was still a virgin. She did not believe in divorce, but took the extraordinary step of turning to the law and having her marriage annulled on the grounds of non-consummation."

In the next appendix, we'll look more deeply into what Marie Stopes and her fellow birth controllers thought of women they considered 'unfit' to be mothers. But this is as good a point as any to explain why elite and highly educated women with busy careers and active social lives had such an ardent zeal to squelch the birthrates of women they considered their inferiors.

The reason lies in the practical application of eugenics. Progress in an evolutionary sense belongs to the group that has the most surviving offspring. In the early years of the twentieth century, groups that regard themselves as the best that humanity had to offer—scientists, socialists, liberals, and members of the establishment—became alarmed at the wide gap between their own birthrates and those of immigrants and the under class.

Two solutions were possible. Elite groups could have more children (positive eugenics) or 'inferior' groups could be forced to have fewer children (negative eugenics). Having less to lose, men were divided over which to emphasize. For rather obvious reasons, women such as Marie Stopes wanted the stress placed on forcing down the birthrates of 'unfit' women. (In America, Margaret Sanger was even more adamant that the solution could not involve a "cradle race" between fit and unfit.)

Present-day debates over issues such as abortion are rooted in this era. Feminist alarm about "forced motherhood" was born in a time when some elite men really did want to force women of their social class to have larger families.

# H

# *Birth Control News* and the 'Unfit'

**Editor:** To avoid giving alarm to their target populations, birth controllers and eugenists were usually vague about who they meant when they mentioned the term 'unfit.' But from time to time they would let slip their real agenda and whose birthrate they wanted to curtail. The second article in this appendix is particularly revealing since the author's "make no difference" remark hints at something that would later become one of the birth controllers' cleverest techniques—inducing target populations to separate sex from having children and to become obsessed with the former. (In this appendix I have made some text bold for emphasis.)

## C.B.C.

### Society for Constructive Birth Control and Racial Progress

#### Dr. Saleeby's Lecture:

For the June General Meeting, the Society had the pleasure of hearing a lecture by Dr. C. W. Saleeby, M.D. Edin, F.R.S. Edin., F.Z.S., on "Birth Control and Eugenics—My Hopes and Fears.". . . .

He then drew attention to the steady decline of the percentage of Anglo-Saxon stock in the United States, and the grave figures discussed in MacDougals's [McDougall's] little book, *National Welfare and National Decay*.

He touched with humor upon the question of whether small families in the Anglo-Saxon Americans were volitional or not, saying: "Mr. Pell may be here to dispute with me, but no one else it likely to question whether the low birth-rate is volitional."

**At the same time in the States, certain low-grade races such, for instance, as the Southern Italians, have an extremely high birth-rate,** which, the lecturer maintained, formed too high a proportion of the American population....

The lecturer deplored the extremely low birthrate in certain distinct classes in our community, in particular the fact that the lowest birth-rate of all is found in just such families as those of Naval and Military Officers, Doctors, Clergy, Solicitors, and persons of independent means, this birth-rate being in general so low as not to maintain the class, and therefore persistently to lead to the extinction of just those qualities which were of such use in any community.

—*Birth Control News,* 1, No. 3 (July 1922), 2.

———— **Editor's Note** ————

Each age tends to have its characteristic bigotry. In the article that follows, English "defective boys" are described as mentally childlike "in everything but stature and carnal appetites," lacking self-control and prone to violence. That was precisely how young black men were being portrayed in American films such as D. W. Griffith's pioneering film, *The Birth of a Nation* (1915).

## DEFECTIVES AND BIRTH CONTROL

*by A. C. B. Teacher of Defectives*

. . . . Life to-day is too strenuous for defectives. The battle is too fierce, the conditions of effective fighting too onerous for any but the mentally fit. Defective boys will never mentally "grow up," but must remain, to the end of their days, in everything but stature and carnal appetites, children.

Are children to be allowed to reproduce? Lastly, it must never be forgotten that, however far the intelligence of a defective may be developed, *yet lack of power to control any one of his lower emotions, "in the heat of the moment," may place him on the gallows!*

**Let the State give every defective full knowledge of how to avoid children, and supply them free with contraceptives; better still, offer them a substantial monetary reward to consent to sterilization, and as long as they are given to understand that the operation will "make no difference," they will in most cases agree.**

We must breed for quality, not quantity in the human species, or the end is nigh.

—*Birth Control News,* 1, No. 4 (Aug. 1922, 3.
(Italics by *Birth Control News*)

———— **Editor's Note** ————

Foreigners as inferior (particularly those from Eastern Europe), the local population mongrelized by race mixing, other races described as "rats" with tribal instincts such as "cunning, bloodthirsty, and cowardly," and Northern European races as superior—all these attitudes were common among birth controllers and chillingly all were soon to be part of the rhetoric of Nazi anti-Semitism.

## CORRESPONDENCE COLUMN

### Practical Experience

Dear Sir,—I have just been sorting out the least good of my young Exmoor ewes to sell them out of the flock, and to keep only the best ewes for breeding from.

The rams are selected twice over, first as lambs and then as yearlings, one ram being enough for 50 ewes.

**This brings home to one the importance of your work in C.B.C. and of lessening the number of births from inferior human parents, and those unable to maintain their children, in our islands.**

**But one feels that unless at the same time the influx of low-caste foreigners, especially from Eastern Europe, is checked, they will fill up the gaps and mongrelize our English and Scotch stock.**

**Like the rats, these low-caste foreigners have large families, and are industrious workers and have strong tribal instincts, but compared with our people they are cunning, bloodthirsty and cowardly.**

In Eastern Europe and Asia large families are the insurance for old age. The more children they have the better will the parents be supported, it being considered the old and elderly parents right to live on their children. So, instead of saving, they have as many children as possible. Their priests encourage them as it increases the number of their "faithful."

**Very likely the Society has this race question in hand.** A few strains of strong races like the French (as the Huguenots) or Scandinavians do good in a pedigree and blend well with us, but these Eastern and Eastern Europe races are altogether inferior to ours.

The London working-classes are distinctly more foreign-looking than they were in my childhood, and we are losing our advantage of living on an island.

I spent several years in various parts of the globe.

Yours sincerely,

C. B. S. Mildmay

October 16, 1922

—*Birth Control News,* 1, No. 7 (Nov. 1922), 2.

——————— **Editor's Note** ———————

The English and Americans were not the only ones to get caught up in this fashionable scientific bigotry. In the next article a Swedish professor agrees with his English counterparts. Note how quickly his views win him an office in the upper ranks of the C.B.C.

### PROSPERITY BAD FOR RACE SAYS SWEDE TEACHER

#### Says Middle Classes Tend to Vanish in Favour of "Human Trash"

Stockholm—Prosperity is a bad thing for a race, because it is one of the first steps toward degeneration, in the opinion of Professor Herman Lundborg, head of the Race Biological Institute at Upsala, Sweden, and well know for his investigations into eugenics and racial biology. . . .

**The professor approves of birth control, however, especially among the less desirable elements of a country's population, and he believes that the reproduction of a race should be limited, as far as possible, to those who are "well born."** He is stoutly opposed to the mixing of races, and gives this as one of the reasons why populations in larger cities tend to degenerate, except in so far as

racially replenished from the countryside districts.

"Europe is decaying, not only as a result of political cataclysms, but also because of a misconception of racial hygiene and a failure to counteract the forces of degeneration."

—*Birth Control News,* 1, No. 8 (Dec. 1922), 1.

### GREAT SWEDISH PROFESSOR

#### Becomes Vice-President of C.B.C

Prof. D. H. Lundborg of the State Institute of Race Biology, Upsala, has been appointed a Vice-President of the C.B.C., and stated, "It makes me glad to hear that you strive for the improvement of the race. With pleasure I become a member of your English Association."

—*Birth Control News,* 1, No. 10 (Feb. 1923), 1.

### C.B.C.

#### Society for Constructive Birth Control and Racial Progress

#### Dr. Hawthorne's Lecture

For the December General Meeting, the Society had the pleasure of hearing a lecture by Dr. Jane Hawthorne on "Birth Control as it affects the Working Mother." . . . .

"Now let me give you the history of another and a different tragedy.

"This was the case of a young, pretty and ambitious girl who fell in love with a young man, whose parents would not allow the marriage because it was thought there was consumption in the family. But like so many other young couples they would not wait, and after a few months of marriage the babies arrived as quickly as it was possible, and after a time the young parents showed unmistakable signs of that dreaded disease—consumption. At the present moment the entire family—father, mother and their children—are attending the tuberculosis dispensary. The husband says when he is remonstrated with that it does him good to lead a married life, and the mother declares she is never better than when she is carrying. **Well, I sup-**

pose men and women have a right over their own lives, but surely they have not the right to bring into the world a brood of diseased children who will not only propagate their kind, but a worse kind!"

—*Birth Control News*, 1, No. 9 (Jan. 1923) 2.

## OTHER SOCIETIES

### Eugenics Education Society

Prof. McBride, lecturing on "Mental Defect and its Inheritability," said that "Any hope of curing defective children by ameliorative and environmental measures was an illusion." "They breed recklessly and are parents of a large amount of our slum population. If the slums were cleared and these people put in ideally constructed dwellings they would reproduce the slums again in thirty or forty years." **"In my opinion there is only one remedy for this state of affairs, and that is the ruthless sterilization of the mental defective so that they may not be able to hand on these defects to posterity."**

—*Birth Control News*, 1, No. 10 (Feb. 1923), 4.

## THE RACIAL PROBLEM IN SCOTLAND

*By the Rev. Duncan Cameron*

The Scottish people have just awakened to the fact that there is a racial problem of the most formidable description in their midst. For many years—indeed, from the beginning of railway construction—there has been a constant stream of emigration of peasantry from Ireland into the industrial areas of Scotland, until in the most populous centres the Irish people now constitute one-fourth, and in some counties—for example Lanarkshire—will soon constitute one-third of the whole population. Contemporary with this invasion of the Irish race, there has been a continuous emigration of Scots to other lands, and this emigration grows larger as time passes....

Englishmen as well as Scotsmen are bound to view these portentous facts with deep concern. The racial conditions in Scotland bode ill for the United Kingdom

and for the Empire. **The Scots—one of the sturdiest and most efficient of races—are gradually leaving their native land, and another people of a different mentality and a lesser calibre are taking their place....**

The Irish Priesthood urge upon their people the duty of large families. This is the way to possess Scotland, the genuine Scot will not be converted to Rome; he must therefore be dispossessed by economic pressure.... And all the great ships sail south and west, carrying away from Scotland the finest peasantry and the most skilled artisans in the world, and leaving behind them such a people.

Sunt Lacrymae Verum [These are true tears]

—*Birth Control News*, 2 No. 5 (Sept. 1923), 3.

## LADY LAYLAND-BARRATT REPLIES TO THE "ANTIS"

In a letter to the Press Lady Layland-Barratt says:—

"It is common knowledge that the British Isles are densely over-populated; that even with the most careful husbandry and the most rigorous rationing of the population, internal resources can only provide food for the people for three months.

"Under these circumstances, what is the use of building more houses? Common sense dictates the urgent necessity for the decrease of the population rather than its extension. **Secondly, how can we expect to be anything but a C3 nation when lunatics, epileptics, dipsomaniacs, feeble-minded, and tubercular persons are permitted to reproduce themselves?** ...

"Until the physically unfit are prevented [from] reproducing themselves, and the working classes restrict the numbers of their children (as the more educated classes do) to the capacity of the mother's physical strength to produce a few healthy offspring and the father's economic capacity to keep, educate, and put them out in the world, the miseries of overcrowded homes will continue."

—*Birth Control News,* 1, No. 8 (Dec. 1922), 2.

## BOOK REVIEWS

Lothrop Stoddard, *The Revolt against Civilization: The Menace of the Underman* . . .

This book is written in a lively style, which makes it easy to read. Much good sense and knowledge lie behind its author's opinions, and as the problems which interest us are here considered from the point of view which even politicians must recognize as coming into their province, we hope that it will be widely read.

To encourage our readers to get the book, we will let the author speak for himself. . .

**"If, then, society is ever to rid itself of its worse burdens, social reform must be increasingly supplemented by racial reform. Unfit individuals as well as unjust social conditions must be eliminated.**

**"Even those persons who carry taints which make parenthood inadvisable need not be debarred from marriage. The sole limitation would be that they should have no children.** And this will be perfectly feasible because, when public opinion acquires the racial viewpoint, the present silly and vicious attitude toward birth control will be abandoned, and undesirable children will not be conceived."

The author is a whole-hearted Constructive Birth Controller:—

"Our particular job is stopping the prodigious spread of inferiority which is now going on. We may be losing our best-stocks, but we are losing them much more slowly than we are multiplying our worst. Our study of differential birth-rates showed us that if these remain unchanged our most intelligent stocks will diminish from one-third to two-thirds in the next hundred years; it also showed that our least intelligent stocks will increase from

six- to ten-fold in the same time. **Obviously, it is this prodigious spawning of inferiors which must at all costs be prevented if society is to be saved from disruption and dissolution. Race cleansing is apparently the only thing that can stop it. Therefore, race cleansing must be our first concern."**

—*Birth Control News,* 1, No. 9 (Jan. 1923), 4

——————— **Editor's Note** ———————

Some of the smugness of the birth controllers is illustrated by their reprinting, in its entirety, a letter by a Roman Catholic leader. The Very Rev. McNabb was, of course, right. With the spread of birth control, welfare programs did begin to see the poor as a group to be eliminated rather than assisted.

## THE ENDOWMENT OF SIN

### (An Open Letter to the Minister of Health)

*By Very Rev. Vincet McNabb, O. P.*

The following open letter is reprinted from the *Catholic Times* of 28th February:

. . . . It is, then, with deep anxiety that we have heard of your intention to aid and develop still more those recent institutions, the Welfare Centres and the like. We are asking ourselves whether the genius loci, the ill-genius of the place, has not already hustled or wheedled you into what will soon show itself to be an ethical inconsequence. . . .

Now, Sir, it is some years since we ventured to prophesy the coming of what Father Martindale already sees in our midst. The Welfare Centres and other organizations were pathetically designed as First Aid against a system which Pope Leo XIII had bid us remedy, and, indeed, remedy quickly. But these First Aid institutions, is so administered as to perpetuate the disease, would be aiding the disease rather than the remedy. Those of us who have lived long within the nidus of the disease realized, as if by intuition, that sooner or later these Welfare and kindred centres would move from an economic to an ethical basis. **Such a move would mean that their end would contradict their beginning. Having been begun in order to aid the poor, they would end with eugenic schemes to eliminate the poor.** Our intuition was so strong that in a spirit of prophecy we dared to say that these Welfare Centres, so peculiar to English-speaking Protestant countries, would be homes of Neo-Malthusianism.

—*Birth Control News,* 2, No. 12 (Apr. 1924), 3.

## C3 POPULATION OF TO-DAY

Dean Inge presided at a Chadwick public lecture on "Some Causes of a 'C3' Population" Given by Professor McBride at the Royal Society of Arts on Friday, May 9.

Dean Inge, in introducing the lecturer, said that it could not be too strongly impressed upon the public mind that the welfare of a nation depended not on its exports and imports, nor on the reduction of debt: it depended on the quality of the men and women that country was turning out. The quality of a population depended partly on environment and partly on heredity. Some people said that to study heredity was absurd, others said that it was contrary to religion, while some said that its effects were already known to all. Nowhere could they find a more uncompromising recognition of eugenics than in the Sermon on the Mount. "The good tree cannot bring forth evil fruit, neither can a corrupt tree bring forth good fruit." If these words did not sanction a reverent and scientific investigation of the problems of heredity, he did not know what words meant.

Professor McBride defined the phrase **"C3 population" as a term to denote the mentally and physically deficient. A "C3" population was largely the result of reckless and improvident reproduction, and the question of the prevention of the appearance of this population and of overpopulation ultimately resolved themselves into the practica-bility of birth control and eventually of sterilization of the unfit.**

"Natural selection," he said, "is the broom by means of which nature keeps other races of animals healthy and clean. If we interfere with its operation in the case of man we must replace it with some other kind of selection; otherwise we shall always be troubled with a C3 population."

"Modern philanthropy, by endeavouring to keep all babies, however feeble, alive, has only accentuated the evil."

—*Birth Control News,* 3, No. 2 (June 1924). 4.

——————— **Editor's Note** ———————

One final remark. It is true that I have, in general, selected the nastier remarks of birth controllers and eugenists for these appendices, while, leaving out the bulk of their oft-repeated claims of concern for poor families and overburdened mothers. But that is of little importance.

Think for a moment of a con man who woes a poor widow with the claim that his scheme will make her old age less grim. Privately, however, he tells his pals that he considers her a fool and intends to leave her penniless. Which of his words are true? Obviously, the second.

Much the same reasoning applies to birth controllers and eugenists. If their vile remarks were true, then their other words are easily explained as deceptions and self-rationalizations. But if the expression of concern were true, their other remarks would have never been made.

# APPENDIX

# I

# *Birth Control News* and Eugenics

---

**Editor:** This appendix explores what *Birth Control News* had to say about eugenics and similar topics during the time frame of Chesterton's book. The first series of articles focus on global population control. Note Bertrand Russell's call for an international authority to "keep the coloured races out" of Europe, Australia and other white areas. Only rarely would birth controllers be so open about this part of their agenda. His call anticipated the present policies of many European countries, policies that stress pro-natalist policies at home, but fund anti-natalist 'family planning' in Africa and Asia. It helps to remember that the "higher and more intelligent communities" of Europe (with their relatively low birthrates) thrust the world into two world wars in the twentieth century. (I have made some text bold for emphasis.)

---

### INTERNATIONAL ASPECTS OF BIRTH CONTROL

. . . . The day will come (if the world does not before that time crash into ruin) when **the higher and more intelligent communities** will make sure that no danger spot, no infectious sore of over-procreation ignorance continues spawning into miserable multitudes in any quarter of the globe to risk the re-infection of the world by such an epidemic of insanity as war.
—*Birth Control News*, 1 No. 3 (July 1922), 3.

### LECTURE BY THE HON. BERTRAND RUSSELL

Dr. Marie Stopes presided over the General Meeting of the C.B.C. on November 16 at the Essex Hall, when the Hon. Bertrand Russell spoke on "Birth Control and International Relations."

. . . . Mr. Russell then described the industry of the Japanese race, but their disadvantage in the lack of raw material, principally iron, for the working of their factories, their effort to procure it from China, which results in the Imperialistic attitude, their persistent efforts to emigrate to Australia, and the equally persistent resolution to keep them out, but the almost impossibility of continuing to isolate a large portion of the world and to say that the coloured races shall not settle there. Sooner or later the pressure will be too great and the barriers will break down. "This policy may last some time, but in the end under it we shall have to give way—we are only putting off the evil day; the one real remedy is birth control, that is getting the people of the world to limit themselves to those numbers which they can keep upon their own soil. . . . **I do not see how we can hope permanently to be strong enough to keep the coloured races out; sooner or later they are bound to overflow, so the best we can do is to hope that those nations will see the wisdom of Birth Control. . . . We need a strong international authority."**

—*Birth Control News,* 1, No. 8 (Dec. 1922), 2.

———— Editor's Note ————

A little over a year later, Bertrand Russell clarified what his "international authority" would do, including controlling births and drugging a population so it displayed the proper "disposition." The Fabian Society to which he spoke these words provides, then and now, much of the intellectual leadership for the British left and the Labor party.

### THE FABIANS AND BIRTH CONTROL

**Mr. Bertrand Russell on Birth Control**
A startling forecast as to the possible effect of the employment of artificial birth control was made by Mr. Bertrand Russell in a Fabian Society lecture at King's Hall, London. Mr. Russell spoke of the production by industrialism of a single world-wide political organism, and said that **if a world organization, however oppressive, were once created ordered progress would again become possible.** Biological sciences had not so far much effect, but might hereafter, through the study of heredity, revolutionize agriculture, make eugenics an exact science, and facilitate artificial determination of sex in children, which would entail profound social changes. Mr. Russell said that so far artificial birth control was probably the most important of their effects. **The dependence of emotional disposition upon the ductless glands, said Mr. Rus-**

**sell, was a discovery of great importance, which would in time make it possible to produce artificially any disposition desired by Governments.** Like all other scientific discoveries, it would have good or bad effects according to the passions of the dominant classes or nations.

—*Birth Control News,* 2, No. 10, (Feb. 1924), 4.

———— Editor's Note ————

The next two articles reveal the scorn birth controllers and their allies had for "reactionary" Chesterton.

### A FOE OF PROGRESS
### G. K. Chesterton, Esq.

No greater contrast could be found to the slender spiritual woman, the friend of progress to whom we have just done homage [Lady Constance Lytton], than the massive, materially-minded man whose intellectual caperings have afforded laughter for five continents.

**Often absurd, sometimes funny, he became ridiculous in his latest book, supposed to be an exposure of Eugenics.** But the professional mountebank should not endeavour to mix with intellectuals, although, I dare say, the misunderstandings of their science which afforded the basis for his book were for him a short cut to a fresh notoriety.

**His tendency is reactionary, and as he succeeds in making most people laugh, his influence in the wrong direction is considerable.** It would be much less than it is were people grounded on the basis of essential scientific truths which he so often juggles with in order to cheat them into following his argument into the ditch or gutter.

Now and then G. K. Chesterton says a true thing, and says it tellingly and well, but the general tendency of his writings is to deal ribaldly with the sacred aspirations of human beings to spiritualize their lives and to mould matter through mind.

Mr. Chesterton is all matter—but he does not matter.

—*Birth Control News,* 1, No. 6, (Oct. 1922), 3.

## MR. CHESTERTON

Dear Editor,—May I say a word as to your paragraph on my cousin, Mr. G. K. Chesterton, of whom I am publishing shortly a book? I do most strongly disagree with his ideas of eugenics. I do not consider he has any good knowledge of the miseries of large families on small purses. I am a member of the C.B.C., and I feel that a great part of the reason of G. K. C.'s attitude is dictated by that arrogant fraudulent Church to which he belongs. He allows his mind to be influenced by a celibate priesthood that is usually immoral, and does not marry because it prefers to be thought not of this world. I wish the Society good luck.

Yours truly

Patrick Braybrooke

Author of *Gilbert Keith Chesterton,* etc.

46, Russell Square, W.C.1

—*Birth Control News,* 1, No. 7 (Nov. 1922), 2.

——— **Editor's Note** ———

The next four articles give some indication how birth controllers intended to interfere with marriage.

## WITHIN AND AROUND,
### Sayings of the Month
*Collected by Samkins Browne*

Mr. E. F. L. Henson, in a letter in the *Times:* "To encourage marriage by subsidizing families . . . would only assist to increase still more the birth-rate in the lower strata of society where birth control is most needed."

—*Birth Control News,* 1, No. 7, (Nov. 1922), 3.

## IN THE U.S.A
### Severe Bill on Eugenics Likely to Pass in Oregon
*(By Special Cable to The Tribune)*

Portland.—Physical and mental tests to determine whether couples should be permitted to marry will be required by the State of Oregon if the Bill introduced in the House becomes a law. Prospects of the Bill passing the legislature are good, and Representative Mrs. Simmons, its author, is confident of it.

**Under its terms, all applicants for marriage licenses must prove that they have at least the mentality of a child of twelve years and be free from communicable or contagious diseases. Failure to pass the examination would preclude the issuance of a license to marry unless one or both of the parties submitted to sterilization.**

—*Birth Control News,* 1 No. 12, (Apr. 1923), 1.

## GYNAECOLOGIST DEMANDS STERILIZATION OF DEFECTIVES

Dr. R. A. Gibbons, Gynaecologist to the Grosvenor Hospital for Women, speaking at the Westminister and Holborn Division of the British Medical Association after their dinner at the Criterion Restaurant, urged the Sterilization of the Unfit and State Certificates of Marriage.

He spoke of the numbers of insane under our care, saying as "there is an annual increase in these numbers, it is clear that there is urgent need to do all in our power to ameliorate this state of affairs." "We should at least endeavour to secure the advent of only healthy children . . . and prevent the propagation of the mentally defective.

Dr. B. Dunlop agreed, but also put in a claim for the value of birth control to achieve this end.

—*Birth Control News,* 2, No. 1 (May 1923), 1.

## IN U.S.A.
### MEDICAL SUPPORT OF BIRTH CONTROL
#### Birth Control and Limitation of Marriage Urged
#### Dr. Brainard, Los Angeles, at Convention Proposes Drastic Steps to Save the Race

San Francisco, June 22.—**Birth control, limitation of marriage to the fit and** sterilization of the criminal and insane were unqualifiedly endorsed by Dr. H. G. Brainard of Los Angeles, president of the California State Medical Association, in an address to-day at the opening session of the Association's annual convention. The speaker, a psychiatrist and neurologist, discussed 'Eugenics.'

Pointing out that draft statistics indicated that one out of every five persons is physically or mentally unfit, Dr. Brainard declared it is up to the medical profession to work for repeal of laws against birth control and to arouse interest in obtaining proper marriage laws.

Dr. Brainard urged the federal government to scan more closely the immigrants, and reject those found mentally unfit.

"By a combination of all these efforts," he concluded, "I believe it is possible to bring about such changes in coming generations as will prevent our country from perishing in a mire of insanity, degeneracy, poverty, immorality, and crime."

—*Birth Control News,* 2 No. 1, (Aug. 1923), 1.

——————— **Editor's Note** ———————

Last are articles fitting no particular classification. *The Lancet* is published by the British Medical Association and represents the medical mainstream. Today we know the dreadful effects of using X-rays as contraceptives, effects that include the birth of children with severe disabilities.

## CONTROL OF CONCEPTION BY IRRADIATION

*The Lancet,* 16 Sept. 1922, gives an account of a recent paper by Dr. Emmrich Markovitz on the method used by a central X-ray laboratory of Vienna. Medical readers are referred to the *Lancet* and to Dr. Markovitz's own account of the work, but as this is in highly technical language the general reader may like to have it summarized.

The work is based on recent investigations, which show that certain portions of the glands of sex are degenerated by X-rays, while at the same time other important tissues in these glands, the so-called interstitial tissues, are hardly, if at all affected. Hence the new technique suggested for temporary control of conception is to treat with X-rays, first the woman, making her temporarily sterilized for a certain number of months, and then, before her power to conceive returns, temporarily sterilizing the husband.

The great advantage of the suggested treatment would be that, while temporary security from unhealthy conception would be secured, it would not involve permanent sterility.

The method should be a most useful one, and it is to be hoped that the English doctors will follow the Continental lead in this matter and investigate the method so as to make it practicable.

—*Birth Control News,* 1, No. 6 (Oct. 1922), 4.

## TENETS OF THE C.B.C.

(16) In short, we are profoundly and fundamentally a pro-baby organization, in favour of producing the largest possible number of healthy, happy children without detriment to the mother, and with the minimum wastage of infants by premature death. We, therefore, as a Society, regret the relatively small families of those best fitted to care for children. In this connection our motto has been "Babies in the right place," and it is just as much the aim of Constructive Birth Control to secure conception to those married people who are healthy, childless and desire children, as it is to furnish security from conception to those who are racially diseased, already overburdened with children, or in any specific way unfitted for parenthood.

—*Birth Control News,* 1, No. 12 (Apr. 1923), 2.

——————— **Editor's Note** ———————

Apart from infanticide and mass murder, there are three basic techniques to limited the population of unwanted groups: abortion, birth control and sterilization. Historically, groups with a eugenic agenda have had advocated two of the three while loudly (perhaps too loudly) denouncing the third. As we see from this poem and the article that follows it, in the 1920s birth control and sterilization were promoted while abortion was strongly condemned.

## LITTLE STRANGER
### M. P. Marjoram
Oh, little stranger, unborn child of mine,
Still lying safe beneath my throbbing heart.
Flesh of my flesh, bone of my bone—and his
Who, in his fatherhood holds vital part. . . .
—*Birth Control News,* 1, No. 12, (Apr. 1923), 3.

## CORRESPONDENCE COLUMN
### A Frequent Evil
Dear Dr. Stopes,—I am in great trouble, and have been advised to write to you, but first obtained your book, "Wise Parenthood," to glean whether I could obtain help from that. Unfortunately you only deal with pre-conception in that. . . .*

* Editor's Note [*Birth Control News*]
Dr. Marie Stopes ask us to state that though the pathos of such cases may wring her heart, she will not and cannot answer such letters. . . .

That there should be such ignorance, though tragic, is perhaps not surprising owing to the suppression of sound knowledge and understanding of such subjects, and, in addition, to the deliberate confusion accorded by the opponents of sex instruction who are quite aware what they are doing, and disseminate the idea that birth control and abortion are the same or comparable things. Dr. Stopes utterly condemns abortion, as does the C.B.C. Society and this paper.
—*Birth Control News,* 2, No. 4, (Aug. 1923), 4.

——— Editor's Note ———
In the early days of birth control, its supporters ridiculed those who suggested that contraceptives would encourage sexual promiscuity. Today, of course, their counterparts claim that teen promiscuity forces us to make contraceptives widely available. In retrospect, it is obvious who was lying.

## OUR PILLORY
### For Lies in Circulation
Another extraordinary statement appeared in the *Church Times* over the signature of Lawrence Phillips of the Theological College, Lichfield. He said: **"That if the use of contraceptives is permitted it is difficult to see why adultery, fornication, sodomy, incest and self-abuse should be considered immoral."** This abominable misstatement reveals a "difficulty" which is present only in the minds of abnormal persons. The reasons which follow in the *Church Times* as being the reasons why the various vile practices mentioned above are immoral are such feeble and superficial reasons, that it is evident that the real truth and beauty of morality is hidden from the writer.
—*Birth Control News,* 2, No. 1, (May 1923), 3.

——— Editor's Note ———
The italics that follow were provided by *Birth Control News,* and may hint at just how closely they agreed with Leonard Darwin's remarks. Leonard was actually the grandson of Charles Darwin.

## PRODUCING THE UNFIT
### Major Darwin and Race Inferiority
### Progress Hindered by Social Reform
The *Daily News* reported:—
"Major Leonard Darwin, a son of Charles Darwin, attacked social reformers, as understood to-day, in the course of his presidential address at the annual meeting of the Eugenics Education Society in London.

"The subject was 'Mate Selection,' and Major Darwin condemned the effects of social reform as forces encouraging the continuance of race inferiority and deterioration.

"*'Pity for others and solicitude for their liberty,' were mentioned as obstacles in the way of race improvement. . . .*

"*'To secure human progress,' he said, 'the inferior types must be eliminated; and all that should be demanded is that this process should be made as little painful as possible.'*"

—*Birth Control News*, 2, No. 4 (Aug. 1923), 1.

——— Editor's Note ———

Note the academic and social prominence of many of those who supported eugenics, immigration restriction, sterilization and birth control.

## In U.S.A.

### Better Breeding of Humans Urged by Eugenics Board

Chicago.—Selective immigration, sterilization of defectives and control of everything having to do with the reproduction of human beings are among the objects of the Eugenics Committee of the United States, which has just issued its sweeping programme for the betterment of the human race.

The committee comes out frankly for birth control and the most widespread distribution of knowledge concerning the breeding of the species.

President Emeritus Charles W. Eliot of Harvard; Senator Royal S. Copeland, former Health Commissioner of New York; Surgeon General H. S. Cummings, Washington; President Livingston Farrand of Cornell University; Dr. David Starr Jordan, Dr. Ray Liman Wilbur, Dr. Charles E. Sawyer and many other noted educational, medical, and social welfare leaders are on the committee's advisory council.

—*Birth Control News*, 2, No. 6, Oct. 1923), 1.

——— Editor's Note ———

We end with remarks from a Canadian Catholic leader that proved all too accurate though the savagery would occur in Germany rather than England or North America.

### FATHER KEEBLE AND BIRTH CONTROL

The Very Rev. Father M. J. Keeble, O.P.G., national director of the Holy Name Society of American, in a sermon at St. Mary's Cathedral, Hamilton, Canada said:

**"They are trying to breed human beings like cattle, to get thoroughbreds, but if they carry that out the savages of Africa will rule the world, for civilization will have become more savage than they."**

—*Birth Control News*, 3, No. 4 (Aug. 1924), 4.

*EUGENICS AND OTHER EVILS*

evolution 6
existing democracies won't go far enough 13, 145
fad to fashion 120
family history examined 17, 144
favored by rich 11
favors abstract over concrete 14
feeble-minded not to be parent 71, 132
few decide on health of many 46
French strong race 55, 158
Germany strong promoter 121
Greek meaning irrelevant 18
help imperialism 129
heroism as crime 15
House of Commons 145
Huguenot strong race 55, 158
impact of World War I 121
indirect methods 125
insane not to be parent 71, 132
Irish 159
Italians low-grade race 39, 156
lead poisoning 138
leave morals out 128
limited by public opinion 145
low birthrates of professionals 156
low-cast foreigners 55, 157
lower races 48, 126
making exact science 163
marriage not abolished 22, 131
marriage prohibited 43, 147
married by police 131
meanness of 98
medicine should support 165
morality, new 62
natural selection 125, 132
negative 8, 42, 51, 132, 133, 138
negro as barbarous race 95, 129
no protection against experiments 102
no right to be a parent 36, 133
not murder 132
not natural selection 65, 132
one-third of U.S. carriers of unfitness 35, 133
origin of term 128
ought to be destroyed 13
people as rats 55, 157
people as trash 158
people compared to sheep 157
popularized in England 121
population control 162
positive 8, 137
professors strong promoters 121

prosperity bad for race 158
rebellion against 25
religion, new 57, 59, 60, 130, 136, 137
results inconclusive 55
ruthless sterilization of slums 60, 159
Scandinavian strong race 55, 158
Scotland 159
selection for segregation 23, 145
sense of humor important 120
Sermon on the Mount 161
slums 60, 159
social sciences 127
society sterilizes fit 151
some critical of birth control 154
spawning of inferiors 98, 160
sterilized defectives 151
studfarm 20
supported by influential 5
Sweden 158
treat like insane 35
tuberculosis (consumption) 158
U.S. more advanced 134
unsuitable marriages 16, 130
use of public schools 144
use of special schools 148
utopia of 124
venereal disease 138
venereal diseased not parent 71, 132
voted on at Oxford 143
welfare centres get rid of poor 161
*Eugenics* (Valere Fallon) 140
*Eugenics and Other Evils* (G. K. Chesterton) 58, 136, 139, 140, 163, 164
Eugenics Committee of the United States 85, 167
Eugenics Education Society 146, 147, 159, 166
Eugenics Record Office 35, 133
*Eugenics Review* 139
"Eugenics: Its Definition, Scope, and Aims" (Francis Galton) 127
Eugenists
    as employer 93
    as eunuch-makers 71
    birth controllers link 8
    breeding people like cattle 99
    Chesterton saw as foolish 9
    Darwinian 51, 133
    described 79
    extend lunatic asylum 31
    five types (sects) 19

# F

# G

say they are king 32
  sterilization of 39, 153
Institute of Hygiene, The 153
Institutionalization
  of feeble-minded 43
  of unfit 23, 103, 134, 145
Insurance Act 116
International authority to force birth control
  163
International population control 162
Ireland 74
Irish in Scotland 159
*Isis, The* 143
Italian
  birth rate too high 39, 156
  southern as low-grade race 39, 156
  thought inferior 5
  too high proportion of U.S. population 39,
    156

# J

James II 67
Jansenists 13
Japanese immigration to Australia 162
Jesuits 13
Jews 5, 9
Joan of Arc 68
Job (biblical) 39
John (King) 67
Jordan, David Starr 85, 167
Journalist, honest 5
Jowett, Mr. 146
Judeo-Christian beliefs 126
Julius Caesar 87

# K

Keats, John 38
Keeble, Very Rev. Father M. J. 167
Kidd, Benjamin 136
King John 44
King of Prussia 57
Kipling, Rudyard 68, 114
Kirby, A. H. P. 149
Ku Klux Klan 9

# L

Labor party (British) 6
Lady Macbeth 41
*Lancet, The* 165
Landladies 97
Langdon-Down, R. 147

Laputa 124
Lauder, Harry 120
Laugher better than eugenics 63
Laughlin, Harry H. 152
Laurie, R. Douglas 149
Law
  clumsy about obvious evils 53
  of England 46
  solicitor's role limited 46
  will obey its own nature 22
Layland-Barratt, Lady 159
Lewis, C. S. 6, 143
Libel cases 29
Liberalism 10, 99
Liberty
  produced skepticism 100
  versus health 102
  without equality 110
Lille war crime 122
Liverpool 83
Lock up strong-minded 43
Lodge, Senator Henry Cabot 9
Lodge, Sir Oliver 20
London University 128
London working class more foreign 55, 158
Long Parliament 13
*Looking Backward* (Edward Bellamy) 24
Lord Bacon 130
Lord Cowdray 96
Louvain war crime 122
Loving wife 62
Lucas, E. V. 76
Lucy, Mr. A. B. 152
Lunacy Laws 23, 35, 63
Lundborg, Prof. D. Herman 158
Lytton, Lady Constance 163

# M

Macbeth 63, 85, 88
MacBride, Prof. E. W. 154
Macdonald, George 33
*Macmillan's Magazine* 123
Madame Hubert 98
Madman, see *Insane*
Males, irresponsible 9
Malthus, Thomas 6, 9, 81
*Mankind in the Making* (H. G. Wells) 54
Marjoram, M. P. 166
Markovitz, Dr. Emmrich 165
Marriage 117

CPSIA information can be obtained
at www.ICGtesting.com
Printed in the USA
LVHW021708061220
673492LV00004B/698

9 781587 420023